Spatial Big Data Science

Zhe Jiang · Shashi Shekhar

Spatial Big Data Science

Classification Techniques for Earth
Observation Imagery

 Springer

Zhe Jiang
Department of Computer Science
University of Alabama
Tuscaloosa, AL
USA

Shashi Shekhar
Department of Computer Science
University of Minnesota
Minneapolis, MN
USA

ISBN 978-3-319-86802-8   ISBN 978-3-319-60195-3   (eBook)
DOI 10.1007/978-3-319-60195-3

Printed on acid-free paper

This Springer imprint is published by Springer Nature
The registered company is Springer International Publishing AG
The registered company address is: Gewerbestrasse 11, 6330 Cham, Switzerland

*To those who have generously helped me
during my Ph.D. study.*

—Zhe Jiang

Preface

With the advancement of remote sensing technology, wide usage of GPS devices in vehicles and cell phones, popularity of mobile applications, crowd sourcing, and geographic information systems, as well as cheaper data storage devices, enormous geo-referenced data is being collected from broader disciplines ranging from business to science and engineering. The volume, velocity, and variety of such geo-reference data are exceeding the capability of traditional spatial computing platform (also called *Spatial big data* or SBD). Emerging spatial big data has transformative potential in solving many grand societal challenges such as water resource management, food security, disaster response, and transportation. However, significant computational challenges exist in analyzing SBD due to the unique spatial characteristics including spatial autocorrelation, anisotropy, heterogeneity, multiple scales, and resolutions. This book discusses the current techniques for spatial big data science, with a particular focus on classification techniques for earth observation imagery big data. Specifically, we introduce several recent spatial classification techniques such as spatial decision trees and spatial ensemble learning to illustrate how to address some of the above computational challenges. Several potential future research directions are also discussed.

Tuscaloosa, USA   Zhe Jiang
Minneapolis, USA   Shashi Shekhar
April 2017

Acknowledgements

This book is based on the doctoral dissertation of Dr. Zhe Jiang under the supervision of Prof. Shashi Shekhar. We would like to thank our collaborator Dr. Joseph Knight and Dr. Jennifer Corcoran from the remote sensing laboratory at the University of Minnesota. Some of the materials are based on a survey collaborated with the members of the spatial computing research group at the University of Minnesota including Reem Ali, Emre Eftelioglu, Xun Tang, Viswanath Gunturi, and Xun Zhou. We would like to acknowledge their collaboration.

Contents

Acronyms

Below is a list of acronyms used in the book.

CAR   Conditional Autoregressive Regression
CART   Classification and Regression Tree
CCA   Canonical Correlation Analysis
CSR   Complete Spatial Randomness
DT   Decision Tree
EM   Expectation and Maximization
EOF   Empirical Orthogonal Functions
ESA   European Space Agency
FTSDT   Focal-Test-Based Spatial Decision Tree
GIS   Geographic Information System
GPU   Graphics Processing Unit
GWR   Geographically Weighted Regression
KDE   Kernel Density Estimation
KMR   K Main Route
LiDAR   Light Detection and Ranging
LISA   Local Indicator of Spatial Association
LTDT   Local-Test-Based Decision Tree
MAUP   Modifiable Area Unit Problem
MODIS   Moderate Resolution Imaging Spectroradiometer
MRF   Markov Random Field
NASA   National Aeronautics and Space Administration
SAR   Spatial Autoregressive Regression
SBD   Spatial Big Data
SDT   Spatial Decision Tree
SEL   Spatial Ensemble Learning
SIG   Spatial Information Gain
SST   Spatial and Spatiotemporal
TAG   Time Aggregate Graph
TEG   Time Expanded Graph
USGS   United States Geological Survey

Part I
Overview of Spatial Big Data Science

Chapter 1
Spatial Big Data

Abstract This chapter discusses the concept of spatial big data, as well as its applications and technical challenges. Spatial big data (SBD), e.g., earth observation imagery, GPS trajectories, temporally detailed road networks, refers to geo-referenced data whose volume, velocity, and variety exceed the capability of current spatial computing platforms. SBD has the potential to transform our society. Vehicle GPS trajectories together with engine measurement data provide a new way to recommend environmentally friendly routes. Satellite and airborne earth observation imagery plays a crucial role in hurricane tracking, crop yield prediction, and global water management. The potential value of earth observation data is so significant that the White House recently declared that full utilization of this data is one of the nation's highest priorities.

1.1 What Is Spatial Big Data?

Traditionally, geospatial data is collected or generated by well-trained experts (e.g., cartographers, census surveyors). The amount of data is usually small. This kind of data can be easily analyzed by visually interpreting patterns on a map. One famous example of analyzing spatial patterns is the Broad Street cholera outbreak [1]. In 1854, a severe outbreak of cholera near the Broad Street of the city of London. At time, people were still not certain on what the causes of the serious disease. Debates were continuing within medical communities on the causes of the persistent outbreak, whether it was by particles in the air or by germ cells ingested through water. The puzzle was solved only after people plotted the disease event instances on a map and found out that hotspots of incidents centered on water pumps (as shown in Fig. 1.1). The deadline cholera was water borne.

Nowadays, however, with the advancement of remote sensors, wide usage of GPS devices in vehicles and cellphones, popularity of mobile applications, crowd sourcing, and geographic information systems, as well as cheap data storage and computational devices, enormous geo-referenced data is being collected from broader disciplines ranging from business to science and engineering, also called *Spatial big data* (SBD) [2]. One example of SBD is geo-social media data. Major social media

© Springer International Publishing AG 2017
Z. Jiang and S. Shekhar, *Spatial Big Data Science*,
DOI 10.1007/978-3-319-60195-3_1

Fig. 1.1 Map of the clusters of cholera cases by John Snow in the London Cholera outbreak of 1854. (image source: Wikipedia)

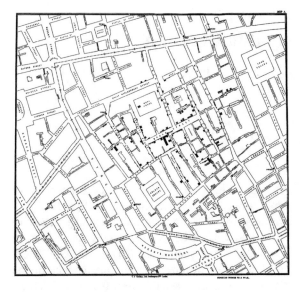

platforms such as Facebook attract billions of active users, most of the users are active on mobile devices such as cellphones, posting their locations via the check-in button. Similarly twitter postings with geo-tags also provide real time "sensor" to monitor major events and locations. Mobile photo-sharing applications such as Instagram collect tens of billions of photos each year. Such a huge multimedia data repository provides detailed content on various objects like famous buildings, parks, and lovely animals, but also provides contextual information via geo-tagging on photos. Another example is earth observation imagery. Remote sensors from satellite and airborne platforms are collecting large volumes of imagery of the earth surface. For instance, MODIS satellites [3] collect imagery of the entire globe every other day. Landsat satellites [4] collect high-resolution image (30 m by 30 m) covering the entire global every sixteen days. NASA itself collects petabytes of earth imagery data each year. Many of the data is free and open in NASA and USGS official websites. Earth observation imagery big data provides unique opportunities for scientists to monitor the dynamics of the earth surface and analyze changes of the land cover types, and to enhance situational awareness for natural disaster management. In transportation, mobile service companies like Uber collects GPS trajectories of vehicles to identify efficient routes and find bottleneck in urban transportation infrastructures. Temporally detailed road network provides traffic volume and speed profile every several minutes each day to provide temporally dynamic route recommendations [5]. Engine measurement data on hundreds of parameters on vehicle speed, acceleration, fuel consumption, emissions and so on, together with GPS trajectories, provide important information on vehicle fuel efficiency and environmental impacts in the real world road network contexts. In public safety, transportation and law enforcement agencies are collecting a large data repositories of traffic accident records and citation records for illegal driving. These rich information provides new opportunities to understand causes of safety issues, and to suggest preventive measures.

Spatial big data can make a difference in several aspects as compared with traditional "smaller" spatial data. At macro level, SBD provides broad spatial coverage of phenomena, making it possible to conduct large scale (global or continental) data analysis. For example, scientists can estimate the amount of global deforestation via Landsat imagery over the last decade. At micro level, SBD also provides high resolution with significant spatial details, making it possible to make "precise" decisions. As an instance of example, high resolution hyperspectral imagery together with GPS promotes the advancement of precision agriculture. Another unique aspect of SBD is that it provides an opportunity to see geographically heterogeneous patterns at different regions. Given the existence of spatial heterogeneity, it is difficult to draw a clear picture of the entire data population unless sufficient data samples are collected. The volume, velocity, and variety of spatial big data, however, exceed the capability of traditional spatial computing platforms. Traditionally, spatial data was analyzed by GIS software tools in the format of flat files (e.g., raster imagery or ESRI shapefiles), or spatial databases (e.g., PostGIS, Oracle Spatial). These tools provide convenient support for basic data processing and analysis. Given the large data volume, quick update rate, and highly heterogeneous nature of spatial big data, these traditional spatial computing platforms become insufficient. For example, Landsat satellites generate earth imagery of the entire globe with 30 m resolution around every sixteen days. Large amount of imagery data is continuously being generated. The portfolio of earth imagery is also diverse with various spatial, spectral, and temporal resolutions.

Spatial big data analytic is the process of discovering interesting, previously unknown, but potentially useful patterns from SBD. Common desired output pattern families include spatial or outliers, associations and tele-connections, predictive models, partitions and summarization, hotspots, as well as change patterns. Spatial outliers are locations whose non-spatial attributes are significantly different from that of spatial neighbors. For example, a house whose size is significantly different from other houses in the same neighborhood is a spatial outlier, even though such a size is not uncommon in the entire city (not a global outlier). Spatial colocation patterns represent types of events that frequently occur close together, such as diseases and certain environment factors. Spatial prediction aims to learn a model that can predict a target response variable (e.g., class labels) based on explanatory features of samples. Examples include classifying earth observation image pixels into different land cover types. Spatial partition focuses on partitioning data into different sub-regions so that data items that are close with each other and similar to each other are in the same sub-region. Summarization aims to provide a compact representation of data, which usually happen after spatial partitioning. Spatial hotspot is an area inside which the intensity of spatial events is higher than outside. For example, downtown area is often the spatial hotspots of crimes in cities. Spatial change patterns represent location or regions where certain non-spatial attributes (e.g., vegetation) change rapidly. Examples include the boundary of different eco-zones such as Sahel, Africa.

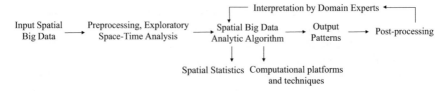

Fig. 1.2 The process of spatial big data science

Figure 1.2 shows the entire process of spatial big data science. It starts with pre-processing of input spatial big data such as noise removal, error correction, geospatial co-registration, map projection, etc. Exploratory data analysis can be done as well to observe data on maps to explore spatial distributions and patterns. After data pre-processing and exploration, spatial big data science algorithms are used to identify useful patterns and to make predictions on the data. These algorithms have spatial statistical foundations for effectiveness and integrate scalable computational techniques and platforms for efficiency. Spatial statistics is unique within the field of statistics in that data samples have spatial dependency instead of being independent and identically distributed. It is commonly studied in the research communities on public health. Spatial computational techniques include data management methods for large scale spatial data such as how to represent, index, and query spatial data. These techniques are special compared with common relational database in that spatial data is often multi-dimensional (e.g., two dimensional objects), and traditional index structures such as B-tree is not applicable. Current spatial computational techniques include multi-dimensional indexing such as R-tree, grid-index, and their variants. The type of input data and the choice of output patterns often determine which kind of algorithms are appropriate to use. After the algorithms produce output spatial patterns, post-processing and pattern interpretations need to be done by domain experts (e.g., wetland expert, criminologist). This step is very important in order to extract real value from spatial big data. Sometimes, domain experts can provide feedback on the output patterns that help refine spatial big data science algorithms, forming a closed loop. Finally, in order to effectively communicate to stake-holders to use the results for decision making, spatial visualization is very important. Geodesign is an example of a set of techniques which integrates the generation of design proposals with simulations on impacts informed by spatial contexts.

1.2 Societal Applications

Spatial big data science are crucial to organizations which make decisions based on large spatial and spatiotemporal datasets, including NASA, the National Geospatial-Intelligence Agency, the National Cancer Institute, the US Department of Transportation, and the National Institute of Justice. These organizations are spread across many application domains.

In earth science and environmental science, researchers need tools to analyze earth observation imagery together with ground in situ field samples to monitor the surface of the planet. This is critically important in various earth science applications including natural resource management (e.g., estimating deforestation in Amazon plain, mapping wetland distribution, monitoring water quantity and quality in open water bodies), disaster management (e.g., flood, forest fires, earthquakes, and landslide), and urbanization studies (e.g., construction and development of urban areas and their environmental impacts). Land cover and land use data product is further used by other simulation models such as hydrological models to provide high-resolution national water forecasting on floods.

In ecology, spatial models have been used to predict the spatial distributions of plant or animal species given environmental factors such as temperature [6, 7]. Empirical (or data driven) models can be compared with models from ecological theories. Ecologists use footprints (spatial polygons) of different endangered species to track areas where more protections are needed. In environment science, spatial prediction methods have been used to interpolate soil properties such as organic matters and top soil thickness [8, 9]. This information is closely related to natural disasters such as landslide.

In public safety, crime analysts are interested in discovering hotspot patterns from crime event records. Given the large data volume, computational tools that automatically detect and visualize hotspot patterns can reduce the burdens of law enforcement agency in decision making, e.g., designing enforcement plans, and allocating police resources. Another similar example is traffic accidents in highways. State agencies are starting to collect the GPS trajectories of their law enforcement vehicles with high frequency (e.g., every 15 s). Such GPS trajectories, together with hotspots of vehicle crash events and driver citation records, provide new opportunities for law enforcement agencies to design police patrol routes that reduce traffic accidents due to illegal driving. Particularly of interests is the potential of predictive analytics that provide suggestions on potential crash event locations so that effective actions can be taken.

In transportation, digital map producers are collecting traffic volume and speed profile on many road segments to provide temporally detailed road networks. Travel time cost at each road segment is estimated every few minutes. GPS trajectories from taxies provide alternative route recommendations based on drivers' experience instead of traditional shortest-path based methods. Logistic companies such as UPS utilize spatial big data such as GPS trajectories and engine measurements as well as driver behaviors to optimize routes, train truck drivers, avoid engine idling time, and reduce unnecessary miles. It is reported that UPS saves millions of gallons of fuel each year [10]. UPS also uses the data for predictive maintenance of their trucks. With the vision of connected vehicles and automatic driving, the amount of data generated from transportation sector and the potential societal value is enormous.

In public health, epidemiologists use spatial big data techniques to plot disease risk map and detect disease outbreak. Previously, due to limited data, disease analysis was often based on aggregated data such as counts in counties. Now with spatial big data, including geo-referenced electronic health records, and environmental data on air quality, it is possible to provide spatially detailed map of disease risk. Moreover, with GPS trajectories of population movement from cellphone records, it is possible to provide more accurate estimation of the spread of transmittable disease. GPS trajectories from mobile apps and local environmental data can also be used for monitoring and alerting for acute disease such as asthma. Predictive models can be constructed to trigger alert when a patient has a high risk to have asthma.

With the emerging themes of automatic driving and Internet of Things, applications of spatial big data will be even broader. The interdisciplinary nature of spatial big data science means that techniques must be developed with awareness of the underlying physics or theories in their application domains [11]. Ignoring domain knowledge and theories, patterns discovered by spatial big data science algorithms may be spurious. For example, climate science studies find that observable predictors for climate phenomena discovered by data driven techniques can be misleading if they do not take into account climate models, locations, and seasons [12]. In this case, statistical significance testing is critically important in order to further validate or discard relationship patterns mined from data. Domain interpretations and comparison of data driven results with results from traditional physical model simulations can also help.

1.3   Challenges

In addition to the huge volume, SBD poses unique statistical and computational challenges due to spatial data characteristics, including spatial autocorrelation, anisotropy, heterogeneity, and multi-scale and resolutions. To address these challenges requires novel data analytic methods.

1.3.1   Implicit Spatial Relationships

Spatial data is often embedded in continuous space, while many classical data mining techniques requires explicit relationships (e.g., transactions in association rule mining), and thus cannot be directly applicable to spatial data. One way to deal with implicit spatial relationships is to materialize the relationships into traditional data input columns and then apply classical big data analytic techniques. For example, in spatial association rule mining, transactions can be created by partitioning the space into a grid. However, the materialization can result in loss of information [13] (e.g., neighboring instances are partitioned into different cells). Moreover, spatial relationships are much more complex than relationship between non-spatial data. For non-spatial data such as numbers or characters, the relationships are relatively simple such as "equal to", "great than", "member of". For spatial data, however, relationships can be defined in difference spaces including set-based space (e.g., union, intersection), topological space (touching, overlap), and metric space (distance, direction). Another issue is the existence of a semantic gap between traditional big data algorithms and spatial and spatiotemporal data. For example, Ring-shaped hotspot pattern is very important in environmental criminology but is hard to characterize in the matrix space as in traditional data mining. Finally, many traditional data mining methods are not spatial or spatiotemporal statistical aware and thus prone to produce spurious spatial patterns. A more preferable way to capture implicit spatial and temporal relationships is to develop statistics and techniques to incorporate spatial and temporal information into the data analytic process.

1.3.2 Spatial Autocorrelation

According to Tobler's first law of geography, "Everything is related to everything else, but near things are more related than distant things." What this law tells us is that spatial data is not statistically independent. Instead, nearby locations tend to resemble each other. This is consistent with our everyday observations. For example, people with similar characteristics, occupation and background tend to live in the same neighborhoods. As another instance of example, land cover classes of nearby pixels in an earth image are often the same due to spatial contiguity of class parcels. In spatial statistics, such spatial dependence is called spatial autocorrelation. Data science techniques that ignore spatial autocorrelation and mistakenly assume an identical and independent distribution (i.i.d.) often generate inaccurate hypotheses or models [13]. For example, many per-pixel classification algorithms such as decision trees and random forests produce salt-and-pepper noise errors in remote sensing image classification. Correcting the errors often involve labor intensive and time consuming post-processing.

1.3.3 Spatial Anisotropy

A second challenge is spatial anisotropy, i.e., the extent of spatial dependency across samples varies across different directions (not isotropic). This is often due to irregular geographical terrains, topographic features and political boundaries. Many current spatial statistics assume isotropy and use spatial neighborhoods with regular shapes (e.g., square window) to model spatial dependency. For example, in Kriging, a popular spatial interpolation method, the covariance between variables at two locations is assumed to be a function of their spatial distance. In other words, data is assumed to be isotropic. This significantly simplifies modeling and parameter estimation, since we can use observations at sample locations to estimate the covariance function. However, this may result in inaccurate models and predictions at the same time. For example, sample observations on river networks are often constrained by the network topological structure and flow directions. Classification and prediction models that assume isotropic spatial dependency and covariance structure in the Euclidean space will be inaccurate. This is critically important in water related applications such as analyzing earth imagery to estimate stream flow volume in hydrology or evaluating water quality in environment science.

1.3.4 Spatial Heterogeneity

Another challenge is the spatial heterogeneity, i.e., spatial data samples do not follow an identical distribution across the entire space. One type of spatial heterogeneity is that samples with the same explanatory features may belong to different class labels in different zones. For example, upland forest looks very similar to wetland forest in spectral values on remote sensing images, but they are from different land cover classes due to different geographical terrains. Another types of spatial heterogeneity is different trends between explanatory variables and response variable in different locations. For instance, in economic studies, it may be possible that old houses are with high price in rural areas, but with low price in urban areas. Though house age is not an effective coefficient for house price when the entire study area is considered, it is an effective coefficient in each local area (rural or urban). In cultural studies, the same body languages or gesture may have different meanings in different countries. These are also called the "spatial" Simpson Paradox. A global model learned from samples in the entire study area may not be effective in different local regions.

1.3.5 Multiple Scales and Resolutions

The last challenge in spatial big data science is that data often exists in multiple spatial scales or resolutions. For example, in earth observation imagery, data resolutions range from sub-meter (high-resolution aerial photos), 30 m (Landsat satellite imagery), and 250 m (MODIS satellite). In precision agriculture, spatial data include both ground sensor observations on soil properties at isolated points and aerial photos on the crop field for the entire area. This poses a challenge since many prediction methods often are developed for spatial data at the same scale or resolution. It is also a great opportunity since spatial data from a single scale or source may have poor quality with noise and missing data, and utilizing data with different scales and resolution can potentially improve the quality as well as spatial and temporal coverage of spatial. Another related data science challenge is that results of spatial analysis depends on the choice of an appropriate spatial scale (e.g., local, regional, global). In spatial statistics, this is also called the modifiable area unit problem (MAUP). For example, spatial autocorrelation values at local level may be significantly different from values at global level, especially when spatial outliers exist. As another instance of example, patterns of spatial interactions between two types of events may be significant in one region of the study area, but insignificant in other areas.

1.4 Organization of the Book

This book overviews spatial big data analytic techniques, with a particular focus on spatial classification methods for earth observation imagery big data. We introduced several recent spatial classification methods in details including spatial decision trees and spatial ensemble learning. Our goal is to provide readers a big picture on spatial big data science, and to illustrate how to address the unique challenges. The organization of the book is as below.

- Chapter 2 provides an overview of current techniques in spatial and spatiotemporal big data science from data mining and computational perspective. Spatial and spatiotemporal (SST) data mining studies the process of discovering interesting, previously unknown, but potentially useful patterns from large SST databases. It has broad application domains including ecology, environmental management, public safety, etc. The complexity of input data (e.g., spatial autocorrelation, anisotropy, heterogeneity) and intrinsic spatial and spatiotemporal relationships limits the usefulness of conventional data mining methods. We review recent computational techniques in SST data mining. This chapter emphasizes the statistical foundation and provides a taxonomy of major pattern families to categorize recent research.
- Chapter 3 overviews earth observation imagery big data from different data sources, including satellites (MODIS, Landsat, Sentinel) and airborne platforms (e.g., LiDAR, Radar, and photogrammetric sensors). It also provides several examples of societal applications where earth imagery classification plays a critical role. The main computational challenges are also discussed that motivate new research. This chapter provides background information for several representative research works in the next three chapters, including spatial information gain-based spatial decision tree, focal-test-based spatial decision tree, and spatial ensemble learning.
- Chapter 4 introduces a novel spatial classification technique called spatial decision trees for geographical classification. Given learning samples from a spatial raster dataset, the geographical classification problem aims to learn a decision tree classifier that minimizes classification errors as well as salt-and-pepper noise. The problem is important in many applications, such as land cover classification in remote sensing and lesion classification in medical diagnosis. However, the problem is challenging due to spatial autocorrelation. Existing decision tree learning algorithms, i.e. ID3, C4.5, CART, produce a lot of salt-and-pepper noise in classification results, due to their assumption that data items are drawn independently from identical distributions. We introduce a spatial decision tree learning algorithm, which incorporates spatial autocorrelation effect by a new spatial information gain (SIG) measure. The proposed approach is evaluated in a case study on a remote sensing dataset from Chanhassen, MN.

- Chapter 5 introduces focal-test-based spatial decision trees that address the challenge of spatial autocorrelation and anisotropy. Given learning samples from a raster dataset, spatial decision tree learning aims to find a decision tree classifier that minimizes classification errors as well as salt-and-pepper noise. The problem has important societal applications such as land cover classification for natural resource management. However, the problem is challenging due to the fact that learning samples show spatial autocorrelation in class labels, instead of being independently identically distributed. Related work relies on local tests (i.e., testing feature information of a location) and cannot adequately model the spatial autocorrelation effect, resulting in salt-and-pepper noise. In contrast, we recently proposed a focal-test-based spatial decision tree (FTSDT), in which the tree traversal direction of a sample is based on both local and focal (neighborhood) information. Preliminary results showed that FTSDT reduces classification errors and salt-and-pepper noise. We also extend our recent work by introducing a new focal test approach with anisotropic spatial neighborhoods that avoids oversmoothing in wedge-shaped areas. We also conduct computational refinement on the FTSDT training algorithm by reusing focal values across candidate thresholds. Theoretical analysis shows that the refined training algorithm is correct and more scalable. Experiment results on real world datasets show that new FTSDT with adaptive neighborhoods improves classification accuracy, and that our computational refinement significantly reduces training time.
- Chapter 6 introduces a novel ensemble learning framework called spatial ensemble to address the challenge of spatial heterogeneity. Given geographical data with class ambiguity, i.e., samples with similar features belonging to different classes in different zones, the spatial ensemble learning (SEL) problem aims to find a decomposition of the geographical area into disjoint zones minimizing class ambiguity and to learn a local classifier in each zone. Class ambiguity is a common issue in many geographical classification applications. For example, in remote sensing image classification, pixels with the same spectral signatures may correspond to different land cover classes in different locations due to heterogeneous geographical terrains. A global classifier may mistakenly classify those ambiguous pixels into one land cover class. However, SEL problem is challenging due to class ambiguity issue, unknown and arbitrary shapes of zonal footprints, and high computational cost due to the potential exponential number of candidate zonal partitions. Related work in ensemble learning either assumes an identical and independent distribution of input data (e.g., bagging, boosting) or decomposes multi-modular input data in the feature vector space (e.g., mixture of experts), and thus cannot effectively decompose the input data in geographical space to reduce class ambiguity. In contrast, we propose a spatial ensemble learning framework that explicitly partition input data in geographical space: first, the input data is preprocessed into homogeneous "patches" via constrained hierarchical spatial clustering; second, patches are grouped into several footprints via greedy seed growing and spatial adjustments. Experimental evaluation on three real world remote sensing datasets show that the proposed approach outperforms related work in classification accuracy.

- Chapter 7 discusses the future research needs in classification of earth observation imagery big data and makes a summary. Most of existing spatial classification methods focus on the challenge of spatial autocorrelation, assuming that data is spatially stationary and isotropic (homogeneous). More research is needed to extend the current methods for spatial data that is heterogeneous and with multiple scales and resolutions. Moreover, with the emergence of geospatial data whose volume, velocity, and variety exceeding traditional spatial computing platforms, scalable classification and prediction algorithms for spatial big data are also needed.

References

1. J. Snow, *On the Mode of Communication of Cholera* (John Churchill, London, 1855), pp. 59–60
2. S. Shekhar, V. Gunturi, M.R. Evans, K. Yang, Spatial big-data challenges intersecting mobility and cloud computing, in *Proceedings of the Eleventh ACM International Workshop on Data Engineering for Wireless and Mobile Access* (ACM, 2012), pp. 1–6
3. NASA, MODIS Moderate Resolution Imaging Spectroradiometer, https://modis.gsfc.nasa.gov/
4. United States Geological Survey, Landsat Missions, https://landsat.usgs.gov/
5. R.Y. Ali, V.M.V. Gunturi, Z. Jiang, S. Shekhar, Emerging applications of spatial network big data in transportation, in *Big Data and Computational Intelligence in Networking* (CRC Press, New York, 2017)
6. M. Austin, Spatial prediction of species distribution: an interface between ecological theory and statistical modelling. Ecol. Model. **157**(2), 101–118 (2002)
7. J. Elith, J.R. Leathwick, Species distribution models: ecological explanation and prediction across space and time. Ann. Rev. Ecol. Evol. Syst. **40**, 677–697 (2009)
8. C.-W. Chang, D.A. Laird, M.J. Mausbach, C.R. Hurburgh, Near-infrared reflectance spectroscopy-principal components regression analyses of soil properties. Soil Sci. Soc. Am. J. **65**(2), 480–490 (2001)
9. T. Hengl, G.B. Heuvelink, A. Stein, A generic framework for spatial prediction of soil variables based on regression-kriging. Geoderma **120**(1), 75–93 (2004)
10. DataFLOQ, Why UPS spends over 1 Billion dollars on Big Data Annually, https://datafloq.com/read/ups-spends-1-billion-big-data-annually/273
11. G. Marcus, E. Davis, Eight (no, nine!) problems with big data. N. Z. Times **6**(04), 2014 (2014)
12. P.M. Caldwell, C.S. Bretherton, M.D. Zelinka, S.A. Klein, B.D. Santer, B.M. Sanderson, Statistical significance of climate sensitivity predictors obtained by data mining. Geophys. Res. Lett. **41**(5), 1803–1808 (2014)
13. S. Shekhar, P. Zhang, Y. Huang, R.R. Vatsavai, Trends in spatial data mining, in *Data Mining: Next Generation Challenges and Future Directions* (2003), pp. 357–380

Chapter 2
Spatial and Spatiotemporal Big Data Science

Abstract This chapter provides an overview of spatial and spatiotemporal big data science. This chapter starts with the unique characteristics of spatial and spatiotemporal data, and their statistical properties. Then, this chapter reviews recent computational techniques and tools in spatial and spatiotemporal data science, focusing on several major pattern families, including spatial and spatiotemporal outliers, spatial and spatiotemporal association and tele-connection, spatial and spatiotemporal prediction, partitioning and summarization, as well as hotspot and change detection.

This chapter overviews the state-of-the-art data mining and data science methods [1] for spatial and spatiotemporal big data. Existing overview tutorials and surveys in spatial and spatiotemporal big data science can be categorized into two groups: early papers in the 1990s without a focus on spatial and spatiotemopral statistical foundations, and recent papers with a focus on statistical foundation. Two early survey papers [2, 3] review spatial data mining from a database approach. Recent papers include brief tutorials on current spatial [4] and spatiotemporal data mining [1] techniques. There are also other relevant book chapters [5–7], as well as survey papers on specific spatial or spatiotemporal data mining tasks such as spatiotemporal clustering [8], spatial outlier detection [9], and spatial and spatiotemporal change footprint detection [10, 11].

This chapter makes the following contributions: (1) We provide a categorization of input spatial and spatiotemporal data types; (2) we provide a summary of spatial and spatiotemporal statistical foundations categorized by different data types; (3) we create a taxonomy of six major output pattern families, including spatial and spatiotemporal outliers, associations and tele-connections, predictive models, partitioning (clustering) and summarization, hotspots, and changes. Within each pattern family, common computational approaches are categorized by the input data types; and (4) we analyze the research trends and future research needs.

Organization of the chapter: This chapter starts with a summary of input spatial and spatiotemporal data (Sect. 2.1) and an overview of statistical foundation (Sect. 2.2). It then describes in detail six main output pattern families including spatial and spatiotemporal outliers, associations and tele-connections, predictive models, partitioning (clustering) and summarization, hotspots, and changes (Sect. 2.3). An

© Springer International Publishing AG 2017

Z. Jiang and S. Shekhar, *Spatial Big Data Science*,

DOI 10.1007/978-3-319-60195-3_2

examination of research trend and future research needs is in Sect. 2.4. Section 2.5 summarizes the chapter.

2.1 Input: Spatial and Spatiotemporal Data

2.1.1 Types of Spatial and Spatiotemporal Data

The data inputs of spatial and spatiotemporal big data science tasks are more complex than the inputs of classical big data science tasks because they include discrete representations of continuous space and time. Table 2.1 gives a taxonomy of different spatial and spatiotemporal data types (or models). Spatial data can be categorized into three models, i.e., the object model, the field model, and the spatial network model [12, 13]. Spatiotemporal data, based on how temporal information is additionally modeled, can be categorized into three types, i.e., temporal snapshot model, temporal change model, and event or process model [14–16]. In the temporal snapshot model, spatial layers of the same theme are time-stamped. For instance, if the spatial layers are points or multi-points, their temporal snapshots are trajectories of points or spatial time series (i.e., variables observed at different times on fixed locations). Similarly, snapshots can represent trajectories of lines and polygons, raster time series, and spatiotemporal networks such as time-expanded graphs (TEGs) and time-aggregated graphs (TEGs) [17, 18]. The temporal change model represents spatiotemporal data with a spatial layer at a given start time together with incremental changes occurring afterward. For instance, it can represent motion (e.g., Brownian motion, random walk [19]) as well as speed and acceleration on spatial points, as well as rotation and deformation on lines and polygons. Event and process models represent temporal information in terms of *events* or *processes*. One way to distinguish events from processes is that events are entities whose properties are possessed timelessly and therefore are not subject to change over time, whereas processes are

Table 2.1 Taxonomy of spatial and spatiotemporal data models

Spatial data	Temporal snapshots (Time series)	Temporal change (Delta/Derivative)	Events/processes
Object model	Trajectories, Spatial time series	Motion, speed, acceleration, split or merge	Spatial or spatiotemporal point process
Field model	Raster time series	Change across raster snapshots	Cellular automation
Spatial network	Spatiotemporal network	Addition or removal of nodes, edges	

entities that are subject to change over time (e.g., a process may be said to be accelerating or slowing down) [20].

2.1.2 Data Attributes and Relationships

There are three distinct types of data attributes for spatiotemporal data, including non-spatiotemporal attributes, spatial attributes, and temporal attributes. Non-spatiotemporal attributes are used to characterize non-contextual features of objects, such as name, population, and unemployment rate for a city. They are the same as the attributes used in the data inputs of classical big data science [21]. Spatial attributes are used to define the spatial location (e.g., longitude and latitude), spatial extent (e.g., area, perimeter) [22, 23], shape, as well as elevation defined in a spatial reference frame. Temporal attributes include the time stamp of a spatial object, a raster layer, or a spatial network snapshot, as well as the duration of a process. Relationships on non-spatial attributes are often explicit, including arithmetic, ordering, and subclass. Relationships on spatial attributes, in contrast, are often implicit, including those in topological space (e.g., meet, within, overlap), set space (e.g., union, intersection), metric space (e.g., distance), and directions. Relationships on spatiotemporal attributes are more sophisticated, as summarized in Table 2.2.

One way to deal with implicit spatiotemporal relationships is to materialize the relationships into traditional data input columns and then apply classical big data science techniques [37–41]. However, the materialization can result in loss of information [7]. The spatial and temporal vagueness which naturally exists in data and relationships usually creates further modeling and processing difficulty in spatial and spatiotemporal big data science. A more preferable way to capture implicit spatial and spatiotemporal relationships is to develop statistics and techniques to incorporate spatial and temporal information into the data science process. These statistics and techniques are the main focus of the survey.

Table 2.2 Relationships on spatiotemporal data

Spatial data	Temporal snapshots (Time series)	Change (Delta/Derivative)	Event/Process
Object model	Spatiotemporal predicates [24], Trajectory distance [25, 26], spatial time series correlation [27], tele-connection [28]	Motion, speed, acceleration, attraction or repulsion, split/merge	Spatiotemporal covariance [19], spatiotemporal coupling for point events, or extended spatial objects [29–34]
Field model	Cubic map algebra [35], temporal correlation, tele-connection	Local, focal, zonal change across snapshots [10]	Cellular automation [36]

2.2   Statistical Foundations

2.2.1   Spatial Statistics for Different Types of Spatial Data

Spatial statistics [19, 42–44] is a branch of statistics concerned with the analysis and modeling of spatial data. The main difference between spatial statistics and classical statistics is that spatial data often fails to meet the assumption of an identical and independent distribution (i.i.d.). As summarized in Table 2.3, spatial statistics can be categorized according to their underlying spatial data type: Geostatistics for point referenced data, lattice statistics for areal data, and spatial point process for spatial point patterns.

Table 2.3 Taxonomy of spatial and spatiotemporal statistics

Spatial model	Spatial statistics	Spatiotemporal statistics
Object model	Geostatistics:	Statistics for spatial time series:
	• Stationarity, isotropy, variograms, Kriging	• Spatiotemporal stationarity, variograms, covariance, Kriging;
	Spatial point processes:	• Temporal autocorrelation, tele-coupling.
	• Poisson point process, spatial scan statistics, Ripley's K-function	Spatiotemporal point processes:
		• Spatiotemporal Poission point process; Spatiotemporal scan statistics; Spatiotemporal K-function.
Field model	Lattice statistics (areal data model):	Statistics for raster time series:
	• W-matrix, spatial autocorrelation, local indicators of spatial association (LISA);	• EOF analysis, CCA analysis;
	• MRF, SAR, CAR, Bayesian hierarchical model	• Spatiotemporal autoregressive model (STAR), Bayesian hierarchical model, dynamic spatiotemporal model (Kalman filter), data assimilation
Spatial network	Spatial network autocorrelation, Network K-function, Network Kriging	

Geostatistics: Geostatistics [44] deal with the analysis of the properties of point reference data, including spatial continuity (i.e., dependence across locations), weak stationarity (i.e., first and second moments do not vary with respect to locations), and isotropy (i.e., uniformity in all directions). For example, under the assumption of weak stationarity (or more specifically intrinsic stationarity), variance of the difference of non-spatial attribute values at two point locations is a function of point location difference regardless of specific point locations. This function is called a variogram [45]. If the variogram only depends on distance between two locations (not varying with respect to directions), it is further called isotropic. Under the assumptions of these properties, Geostatistics also provides a set of statistical tools such as Kriging [45], which can be used to interpolate non-spatial attribute values at unsampled locations. Finally, real-world spatial data may not always satisfy the stationarity assumption. For example, different jurisdictions tend to produce different laws (e.g., speed limit differences between Minnesota and Wisconsin). This effect is called spatial heterogeneity or non-stationarity. Special models (e.g., geographically weighted regression, or GWR [46]) can be further used to model the varying coefficients at different locations.

Lattice statistics: Lattice statistics studies statistics for spatial data in the field (or areal) model. Here a lattice refers to a countable collection of regular or irregular cells in a spatial framework. The range of spatial dependency among cells is reflected by a neighborhood relationship, which can be represented by a contiguity matrix called a W-matrix. A spatial neighborhood relationship can be defined based on spatial adjacency (e.g., rook or queen neighborhoods) or Euclidean distance or, in more general models, cliques and hypergraphs [47]. Based on a W-matrix, spatial autocorrelation statistics can be defined to measure the correlation of a non-spatial attribute across neighboring locations. Common spatial autocorrelation statistics include Moran's I, Getis-Ord $Gi*$, Geary's C, Gamma index Γ [48], as well as their local versions called local indicators of spatial association (LISA) [49]. Several spatial statistical models, including the spatial autoregressive model (SAR), conditional autoregressive model (CAR), Markov random field (MRF), as well as other Bayesian hierarchical models [42], can be used to model lattice data. Another important issue is the modifiable areal unit problem (MAUP) (also called the multi-scale effect) [50], an effect in spatial analysis that results for the same analysis method will change on different aggregation scales. For example, analysis using data aggregated by states will differ from analysis using data at individual family level.

Spatial point processes: A spatial point process is a model for the spatial distribution of the points in a point pattern. It differs from point reference data in that the random variables are locations. Examples include positions of trees in a forest and locations of bird habitats in a wetland. One basic type of point process is a homogeneous spatial Poisson point process (also called complete spatial randomness, or CSR) [19], where point locations are mutually independent with the same intensity over space. However, real-world spatial point processes often show either spatial aggregation (clustering) or spatial inhibition instead of complete spatial independence as in CSR. Spatial statistics such as Ripley's K-function [51, 52], i.e., the average number of points within a certain distance of a given point normalized by the average intensity, can be used to test spatial aggregation of a point pattern against CSR. Moreover, real-world spatial point processes such as crime events often contain hotspot areas instead of following homogeneous intensity across space. A spatial scan statistic [53] can be used to detect these hotspot patterns. It tests whether the intensity of points inside a scanning window is significantly higher (or lower) than outside. Though both the K-function and spatial scan statistics have the same null hypothesis of CSR, their alternative hypotheses are quite different: The K-function tests whether points exhibit spatial aggregation or inhibition instead of independence, while spatial scan statistics assume that points are independent and test whether a local hotspot with much higher intensity than outside exists. Finally, there are other spatial point processes such as the Cox process, in which the intensity function itself is a random function over space, as well as a cluster process, which extends a basic point process with a small cluster centered on each original point [19]. For extended spatial objects such as lines and polygons, spatial point processes can be generalized to line processes and flat processes in stochastic geometry [54].

Spatial network statistics: Most spatial statistics research focuses on the Euclidean space. Spatial statistics on the network space are much less studied. Spatial network space, e.g., river networks and street networks, is important in applications of environmental science and public safety analysis. However, it poses unique challenges including directionality and anisotropy of spatial dependency, connectivity, as well as high computational cost. Statistical properties of random fields on a network are summarized in [55]. Recently, several spatial statistics, such as spatial autocorrelation, K-function, and Kriging, have been generalized to spatial networks [56–58]. Little research has been done on spatiotemporal statistics on the network space.

2.2.2   Spatiotemporal Statistics

Spatiotemporal statistics [19, 59] combine spatial statistics with temporal statistics (time series analysis [60], dynamic models [59]). Table 2.3 summarizes common statistics for different spatiotemporal data types, including spatial time series, spatiotemporal point process, and time series of lattice (areal) data.

Spatial time series: Spatial statistics for point reference data have been generalized for spatiotemporal data [61]. Examples include spatiotemporal stationarity, spatiotemporal covariance, spatiotemporal variograms, and spatiotemporal Kriging [19, 59]. There is also temporal autocorrelation and tele-coupling (high correlation across spatial time series at a long distance). Methods to model spatiotemporal process include physics inspired models (e.g., stochastically differential equations) [19] and hierarchical dynamic spatiotemporal models (e.g., Kalman filtering) for data assimilation [19].

Spatiotemporal point process: A spatiotemporal point process generalizes the spatial point process by incorporating the factor of time. As with spatial point processes, there are spatiotemporal Poisson process, Cox process, and cluster process. There are also corresponding statistical tests including a spatiotemporal K-function and spatiotemporal scan statistics [19].

Time series of lattice (areal) data: Similar to lattice statistics, there are spatial and temporal autocorrelation, SpatioTemporal Autoregressive Regression (STAR) model [62], and Bayesian hierarchical models [42]. Other spatiotemporal statistics include empirical orthogonal function (EOF) analysis (principle component analysis in geophysics), canonical correlation analysis (CCA), and dynamic spatiotemporal models (Kalman filter) for data assimilation [59].

2.3 Output Pattern Families

2.3.1 *Spatial and Spatiotemporal Outlier Detection*

This section reviews techniques for spatial and spatiotemporal outlier detection. The section begins with a definition of spatial or spatiotemporal outliers by comparison with global outliers. Spatial and spatiotemporal outlier detection techniques are summarized according to their input data types.

Problem definition: To understand the meaning of spatial and spatiotemporal outliers, it is useful first to consider global outliers. Global outliers [63, 64] have been informally defined as observations in a dataset which appear to be inconsistent with the remainder of that set of data, or which deviate so much from other observations as to arouse suspicions that they were generated by a different mechanism. In contrast, a spatial outlier [65] is a spatially referenced object whose non-spatial attribute values differ significantly from those of other spatially referenced objects in its spatial neighborhood. Informally, a spatial outlier is a local *instability* or *discontinuity*. For example, a new house in an old neighborhood of a growing metropolitan area is a spatial outlier based on the non-spatial attribute house age. Similarly, a spatiotemporal outlier generalizes spatial outliers with a spatiotemporal neighborhood instead of a spatial neighborhood.

Statistical foundation: The spatial statistics for spatial outlier detection are also applicable to spatiotemporal outliers as long as spatiotemporal neighborhoods are well-defined. The literature provides two kinds of bipartite multi-dimensional tests: graphical tests, including variogram clouds [66] and Moran scatterplots [44, 49], and quantitative tests, including scatterplot [67] and neighborhood spatial statistics [65].

2.3.1.1 Spatial Outlier Detection

The *visualization approach* plots spatial locations on a graph to identify spatial outliers. The common methods are variogram clouds and Moran scatterplot as introduced earlier.

The *neighborhood approach* defines a spatial neighborhood, and a spatial statistic is computed as the difference between the non-spatial attribute of the current location and that of the neighborhood aggregate [65]. Spatial neighborhoods can be identified by distance on spatial attributes (e.g., K-nearest neighbors), or by graph connectivity (e.g., locations on road networks). This work has been extended in a number of ways to allow for multiple non-spatial attributes [68], average and median attribute value [69], weighted spatial outliers [70], categorical spatial outlier [71], local spatial outliers [72], and fast detection algorithms [73], and parallel algorithms on GPU for big spatial event data [74].

2.3.1.2 Spatiotemporal Outlier Detection

The intuition behind spatiotemporal outlier detection is that they reflect "discontinuity" on non-spatiotemporal attributes within a spatiotemporal neighborhood. Approaches can be summarized according to the input data types.

Outliers in spatial time series: For spatial time series (on point reference data, raster data, as well as graph data), basic spatial outlier detection methods, such as visualization-based approaches and neighborhood-based approaches, can be generalized with a definition of spatiotemporal neighborhoods.

Flow Anomalies: Given a set of observations across multiple spatial locations on a spatial network flow, flow anomaly discovery aims to identify dominant time intervals where the fraction of time instants of significantly mismatched sensor readings exceeds the given percentage-threshold. Flow anomaly discovery can be considered as detecting *discontinuities* or *inconsistencies* of a non-spatiotemporal attribute within a neighborhood defined by the flow between nodes, and such discontinuities are persistent over a period of time. A time-scalable technique called SWEET (Smart Window Enumeration and Evaluation of persistent-Thresholds) was proposed [75] that utilizes several algebraic properties in the flow anomaly problem to discover these patterns efficiently.

2.3.2 Spatial and Spatiotemporal Associations, Tele-Connections

This section reviews techniques for identifying spatial and spatiotemporal association as well as tele-connections. The section starts with the basic spatial association (or colocation) pattern and moves on to spatiotemporal association (i.e., spatiotemporal co-occurrence, cascade, and sequential patterns) as well as spatiotemporal tele-connection.

Pattern definition: Spatial association, also known as spatial colocation patterns [76], represents subsets of spatial event types whose instances are often located in close geographic proximity. Real-world examples include symbiotic species, e.g., the Nile Crocodile and Egyptian Plover in ecology. Similarly, spatiotemporal association patterns represent spatiotemporal object types whose instances often occur in close geographic and temporal proximity. Spatiotemporal coupling patterns can be categorized according to whether there exists temporal ordering of object types: spatiotemporal (mixed drove) co-occurrences [77] are used for unordered patterns, spatiotemporal cascades [31] for partially ordered patterns, and spatiotemporal sequential patterns [33] for totally ordered patterns. Spatiotemporal tele-connection [27] represents patterns of significantly positive or negative temporal correlation between a pair of spatial time series.

Challenges: Mining patterns of spatial and spatiotemporal association are challenging due to the following reasons: First, there is no explicit transaction in continuous space and time; second, there is potential for over-counting; and third, the number of candidate patterns is exponential, and a trade-off between statistical rigor of output patterns and computational efficiency has to be made.

Statistical foundation: The underlying statistic for spatiotemporal coupling patterns is the cross-K-function, which generalizes the basic Ripley's K-function (introduced in Sect. 2.2) for multiple event types.

Common approaches: The following subsections categorize common computational approaches for discovering spatial and spatiotemporal couplings by different input data types.

Spatial colocation: Mining colocation patterns can be done via statistical approaches including cross-K-function with Monte Carlo simulation [44], mean nearest neighbor distance, and spatial regression model [78], but these methods are often computationally very expensive due to the exponential number of candidate patterns. In contrast, data mining approaches aim to identify colocation patterns like association rule mining. Within this category, there are transaction-based approaches and distance-based approaches. A transaction-based approach defines transactions over space (e.g., around instances of a reference feature) and then uses an Apriori-like algorithm [79]. A distance-based approach defines a distance-based pattern called k-neighboring class sets [80] or using an event centric model [76] based on a definition of *participation index*, which is an upper bound of cross-K-function statistic and has an anti-monotone property. Recently, approaches have been proposed to identify colocations for extended spatial objects [81] or rare events [82], regional colocation patterns [83–85] (i.e., pattern is significant only in a subregion), statistically significant colocation [86], as well as design fast algorithms [87].

Spatiotemporal event associations represent subsets of two or more event types whose instances are often located in close spatial and temporal proximity. Spatiotempral event associations can be categorized into *spatiotemporal co-occurrences*, *spatiotemporal cascades*, and *spatiotemporal sequential patterns* for temporally unordered events, partially ordered events, and totally ordered events, respectively. To discover spatiotemporal co-occurrences, a monotonic composite interest measure and novel mining algorithms are presented in [77]. A filter-and-refine approach has also been proposed to identify spatiotemporal co-occurrences on extended spatial objects [30]. A spatiotemporal sequential pattern represents a "chain reaction" from different event types. A measure of *sequence index*, which can be interpreted by K-function statistic, was proposed in [33], together with computationally efficient algorithms. For spatiotemporal cascade patterns, a statistically meaningful metric was proposed to quantify interestingness and pruning strategies were proposed to improve computational efficiency [31].

Spatiotemporal association from moving objects trajectories: Mining spatiotemporal association from trajectory data is more challenging than from spatiotemporal event data due to the existence of temporal duration, different moving directions, and imprecise locations. There are a variety of ways to define spatiotemporal association patterns from moving object trajectories. One way is to generalize the definition from spatiotemporal event data. For example, a pattern called spatiotemporal colocation episodes is defined to identify frequent sequences of colocation patterns that share a common event (object) type [88]. As another example, a spatiotemporal sequential pattern is defined based on decomposition of trajectories into line segments and identification of frequent region sequences around the segments [89]. Another way is to define spatiotemporal association as group of objects that frequently move together, either focusing on the footprints of subpaths (region sequences) that are commonly traversed [90] or subsets of objects that frequently move together (also called *travel companion*) [91].

Spatial time series oscillation and tele-connection: Given a collection of spatial time series at different locations, tele-connection discovery aims to identify pairs of spatial time series whose correlation is above a given threshold. Tele-connection patterns are important in understanding oscillations in climate science. Computational challenges arise from the large number of candidate pairs and the length of time series. An efficient index structure, called a cone-tree, as well as a filter-and-refine approach [27], has been proposed which utilizes spatial autocorrelation of nearby spatial time series to filter out redundant pairwise correlation computation. Another challenge is the existence of spurious "high correlation" patterns that happen by chance. Recently, statistical significant tests have been proposed to identify statistically significant tele-connection patterns called dipoles from climate data [28]. The approach uses a "wild bootstrap" to capture the spatiotemporal dependencies and takes account of the spatial autocorrelation, the seasonality, and the trend in the time series over a period of time.

2.3.3   Spatial and Spatiotemporal Prediction

Problem definition: Given training samples with features and a target variable as well as a spatial neighborhood relationship among samples, the problem of *spatial prediction* aims to learn a model that can predict the target variable based on features. What distinguishes spatial prediction from traditional prediction problem in data mining is that data items are embedded in space and often violate the common assumption of an identical and independent distribution (i.i.d.). Spatial prediction problems can be further categorized into *spatial classification* for nominal (i.e., categorical) target variables and *spatial regression* for numeric target variables.

Challenges: The unique challenges of spatial and spatiotemporal prediction come from the special characteristics of spatial and spatiotemporal data, which include spatial and temporal autocorrelation, spatial heterogeneity, and temporal non-stationarity, as well as the multi-scale effect. These unique characteristics violate the common assumption in many traditional prediction techniques that samples follow an identical and independent distribution (i.i.d.). Simply applying traditional prediction techniques without incorporating these unique characteristics may produce hypotheses or models that are inaccurate or inconsistent with the dataset.

Statistical foundations: Spatial and spatiotemporal prediction techniques are developed based on spatial and spatiotemporal statistics, including spatial and temporal autocorrelation, spatial heterogeneity, temporal non-stationarity, and multiple areal unit problem (MAUP) (see Sect. 2.2).

Computational approaches: The following subsections summarize common spatial and spatiotemporal prediction approaches for different data types. We further categorize these approaches according to the challenges that they address, including spatial and spatiotemporal autocorrelation, spatial heterogeneity, spatial multi-scale effect, and temporal non-stationarity, and introduce each category separately below.

2.3.3.1 Spatial Autocorrelation or Dependency

According to Tobler's first law of geography [92], "everything is related to everything else, but near things are more related than distant things." The spatial autocorrelation effect tells us that spatial samples are not statistically independent, and nearby samples tend to resemble each other. There are different ways to incorporate the effect of spatial autocorrelation or dependency into predictive models, including spatial feature creation, explicit model structure modification, and spatial regularization in objective functions.

Spatial feature creation: The main idea is to create new features that incorporate spatial contextual (neighborhood) information. Spatial features can be generated directly from spatial aggregation [93] and indirectly from multi-relationship (or spatial association) rules between spatial entities [94–96] or from spatial transformation of raw features [97]. After spatial features are generated, they can be fed into a general prediction model. One advantage of this approach is that it could utilize many existing predictive models without significant modification. However, spatial feature creation in preprocessing phase is often application specific and time-consuming.

Spatial interpolation: Given observations of a variable at a set of locations (point reference data), spatial interpolation aims to measure the variable value at an unsampled location [98]. These techniques are broadly classified into three categories: geostatistical, non-geostatistical, and some combined approaches. Among the non-geostatistical approaches, the nearest neighbors, inverse distance weighting, etc., are the mostly used techniques in the literature. *Kriging* is the most widely used geostatistical interpolation technique, which represents a family of generalized least-squares regression-based interpolation techniques [99]. *Kriging* can be broadly classified into two categories: univariate (only variable to be predicted) and multivariate (there are some *covariates*, also called explanatory variables). Unlike the non-geostatistical or traditional interpolation techniques, this estimator considers both the distance and the degree of variation between the sampled and unsampled locations for the random variable estimation. Among the univariate kriging methods, the *simple kriging* and *ordinary kriging*, and in multivariate scenario, the *ordinary cokriging, universal kriging* and *kriging with external drift* are the most popular and widely used technique in the study of spatial interpolation [98, 100]. However, the *kriging* suffers from some acute shortcomings of assuming the isotopic nature of the random variables.

Markov random field (MRF): MRF [45] is a widely used model in image classification problems. It assumes that the class label of one pixel only depends on the class labels of its predefined neighbors (also called Markov property). In spatial classification problem, MRF is often integrated with other non-spatial classifiers to incorporate the spatial autocorrelation effect. For example, MRF has been integrated with maximum likelihood classifiers (MLC) to create Markov random field (MRF)-based Bayesian classifiers [101], in order to avoid salt-and-pepper noise in prediction [102]. Another example is the model of Support Vector Random Fields [103].

Spatial Autoregressive Model (SAR): In the spatial autoregression model, the spatial dependencies of the error term, or the dependent variable, are directly modeled in the regression equation [104]. If the dependent values y_i are related to each other, then the regression equation can be modified as $y = \rho W y + X\beta + \epsilon$, where W is the neighborhood relationship contiguity matrix and ρ is a parameter that reflects the strength of the spatial dependencies between the elements of the dependent variable. For spatial classification problems, logistic transformation can be applied to SAR model for binary classes.

Conditional autoregressive model (CAR): In the conditional autoregressive model [45], the spatial autocorrelation effect is explicitly modeled by the conditional probability of the observation of a location given observations of neighbors. CAR is essentially a Markov random field. It is often used as a spatial term in Bayesian hierarchical models.

Spatial accuracy objective function: In traditional classification problems, the objective function (or loss function) often measures the zero-one loss on each sample, no matter how far the predicted class is from the location of the actuals. For example, in bird nest location prediction problem on a rasterized spatial field, a cell's predicted class (e.g., bird nest) is either correct or incorrect. However, if a cell mistakenly predicted as the bird nest class is very close to an actual bird nest cell, the prediction accuracy should not be considered as zero. Thus, spatial accuracy [105, 106] has been proposed to measure not only how accurate each cell is predicted itself but also how far it is from an actual class locations. A case study has shown that learning models based on proposed objective function produce better accuracy in bird nest location prediction problem. Spatial objective function has also been proposed in active learning [107], in which the cost of additional label not only considers accuracy but also travel cost between locations to be labeled.

2.3.3.2   Spatial Heterogeneity

Spatial heterogeneity describes the fact that samples often do not follow an identical distribution in the entire space due to varying geographic features. Thus, a global model for the entire space fails to capture the varying relationships between features and the target variable in different subregion. The problem is essentially the multi-task learning problem, but a key challenge is how to identify different tasks (or regional or local models). Several approaches have been proposed to learn local or regional models. Some approaches first partition the space into homogeneous regions and learn a local model in each region. Others learn local models at each location but add spatial constraint that nearby models have similar parameters.

Geographically Weighted Regression (GWR): One limitation of the spatial autoregressive model (SAR) is that it does not account for the underlying spatial heterogeneity that is natural in the geographic space. Thus, in a SAR model, coefficients β of covariates and the error term ϵ are assumed to be uniform throughout the entire geographic space. One proposed method to account for spatial variation in model parameters and errors is Geographically Weighted Regression [46]. The regression equation of GWR is $y = X\beta(s) + \epsilon(s)$, where $\beta(s)$ and $\epsilon(s)$ represent the spatially parameters and the errors, respectively. GWR has the same structure as standard linear regression, with the exception that the parameters are spatially varying. It also assumes that samples at nearby locations have higher influence on the parameter estimation of a current location. Recently, a multi-model regression approach is proposed to learn a regression model at each location but regularize the parameters to maintain spatial smoothness of parameters at neighboring locations [108].

2.3.3.3   Multi-scale Effect

One main challenge in spatial prediction is the Multiple Area Unit Problem (MAUP), which means that analysis results will vary with different choices of spatial scales. For example, a predictive model that is effective at the county level may perform poorly at states level. Recently, a computation technique has been proposed to learn a predict models from different candidate spatial scales or granularity [94].

2.3.3.4   Spatiotemporal Autocorrelation

Approaches that address the spatiotemporal autocorrelation are often extensions of previously introduced models that address spatial autocorrelation effect by further considering the time dimension. For example, SpatioTemporal Autoregressive Regression (STAR) model [44] extends SAR by further modeling temporal or spatiotemporal dependency across variables at different locations. Spatiotemporal Kriging [59] generalizes spatial kriging with a spatiotemporal covariance matrix and variograms. It can be used to make predictions from incomplete and noisy spatiotemporal data. *Spatiotemporal relational probability trees and forests* [109] extend decision tree classifiers with tree node tests on spatial properties on objects and random field as well as temporal changes. To model spatiotemporal events such as disease counts in different states at a sequence of times, *Bayesian hierarchical models* are often used, which incorporate the spatial and temporal autocorrelation effects in explicit terms.

2.3.3.5 Temporal Non-stationarity

Hierarchical dynamic spatiotemporal models (DSTMs) [59], as the name suggests, aim to model spatiotemporal processes dynamically with a Bayesian hierarchical framework. There are three levels of models in the hierarchy: a data model on the top, a process model in the middle, and a parameter model at the bottom. A data model represents the conditional dependency of (actual or potential) observations on the underlying hidden process with latent variables. A process model captures the spatiotemporal dependency within the process model. A parameter model characterizes the prior distributions of model parameters. DSTMs have been widely used in climate science and environment science, e.g., for simulating population growth or atmospheric and oceanic processes. For model inference, Kalman filter can be used under the assumption of linear and Gaussian models.

2.3.3.6 Prediction for Moving Objects

Mining moving object data such as GPS trajectories and check-in histories has become increasingly important. Due to space limit, we briefly discuss some representative techniques for three main problems: trajectory classification, location prediction, and location recommendation.

Trajectory classification: This problem aims to predict the class of trajectories. Unlike spatial classification problems for spatial point locations, trajectory classification can utilize the order of locations visited by moving objects. An approach has been proposed that uses frequent sequential patterns within trajectories for classification [110].

Location prediction: Given historical locations of a moving object (e.g., GPS trajectories, check-in histories), the location prediction problem aims to forecast the next place that the object will visit. Various approaches have been proposed [111–113]. The main idea is to identify the frequent location sequences visited by moving objects, and then, next location can be predicted by matching the current sequence with historical sequences. Social, temporal, and semantic information can also be incorporated to improve prediction accuracy. Some other approaches use hidden Markov model to capture the transition between different locations. Supervised approaches have also been used.

Location recommendation: Location recommendation [114–118] aims to suggest potentially interesting locations to visitors. Sometimes, it is considered as a special location prediction problem which also utilizes location histories of other moving objects. Several factors are often considered for ranking candidate locations, such as local popularity and user interests. Different factors can be simultaneously incorporated via generative models such as latent Dirichlet allocation (LDA) and probabilistic matrix factorization techniques.

2.3.4 Spatial and Spatiotemporal Partitioning (Clustering) and Summarization

Problem definition: Spatial partitioning aims to divide spatial items (e.g., vector objects, lattice cells) into groups such that items within the same group have high proximity. Spatial partitioning is often called *spatial clustering*. We use the name "spatial partitioning" due to the unique nature of spatial data, i.e., grouping spatial items also mean partitioning the underlying space. Similarly, *spatiotemporal partitioning*, or *spatiotemporal clustering*, aims to group similar spatiotemporal data items and thus partition the underlying space and time. After spatial or spatiotemporal partitioning, one often needs to find a compact representation of items in each partition, e.g., aggregated statistics or representative objects. This process is further called *spatial or spatiotemporal summarization*.

Challenges: The challenges of spatial and spatiotemporal partitioning come from three aspects. First, patterns of spatial partitions in real-world datasets can be of various shapes and sizes and are often mixed with noise and outliers. Second, relationships between spatial and spatiotemporal data items (e.g., polygons, trajectories) are more complicated than traditional non-spatial data. Third, there is a trade-off between quality of partitions and computational efficiency, especially for large datasets.

Computational approaches: Common spatial and spatiotemporal partitioning approaches are summarized in below according to the input data types.

2.3.4.1 Spatial Partitioning (Clustering)

Spatial and spatiotemporal partitioning approaches can be categorized by input data types, including spatial points, spatial time series, trajectories, spatial polygons, raster images, raster time series, spatial networks, and spatiotemporal points.

Spatial point partitioning (clustering): The goal is to partition two-dimensional points into clusters in Euclidean space. Approaches can be categorized into global methods, hierarchical methods, and density-based methods according to the underlying assumptions on the characteristics of clusters [119]. Global methods assume clusters to have "compact" or globular shapes and thus minimize the total distance from points to their cluster centers. These methods include K-means, K-medoids, EM algorithm, CLIQUE, BIRCH, and CLARANS [21]. Hierarchical methods [21] form clusters hierarchically in a top-down or bottom-up manner and are robust to outliers since outliers are often easily separated out. Chameleon [120] is a graph-based hierarchical clustering method that first creates a sparse k-nearest neighbor graph, then partitions the graph into small clusters, and hierarchically merges small clusters whose properties stay mostly unchanged after merging. Density-based methods such as DB-Scan [121] assume clusters to contain dense points and can have arbitrary shapes. When the density of points varies across space, the similarity measure of *shared nearest neighbors* [122] can be used. Voronoi diagram [123] is another space partitioning technique that is widely used in applications of location-based service. Given a set of spatial points in Euclidean space, a Voronoi diagram partitions the space into cells according to the nearest spatial points.

Spatial polygon clustering: Spatial polygon clustering is more challenging than point clustering due to the complexity of distance measures between polygons. Distance measures on polygons can be defined based on dissimilarities on spatial attribute (e.g., Hausdorff distance, ratio of overlap, extent, direction, and topology) as well as non-spatial attributes [124, 125]. Based on these distance measures, traditional point clustering algorithms such as K-means, CLARANS, and shared nearest neighbor algorithm can be applied.

Spatial areal data partitioning: Spatial areal data partitioning has been extensively studied for image segmentation tasks. The goal is to partition areal data (e.g., images) into regions that are homogeneous in non-spatial attributes (e.g., color or gray tone and texture) while maintaining spatial continuity (without small holes). Similar to spatial point clustering, there is no uniform solution. Common approaches can be categorized into non-spatial attribute-guided spatial clustering, single, centroid, or hybrid linkage region growing schemes, and split-and-merge scheme. More details can be found in a survey on image segmentation [126].

Spatial network partitioning: Spatial network partitioning (clustering) is important in many applications such as transportation and VLSI design. Network Voronoi diagram is a simple method to partition spatial network based on common closest interesting nodes (e.g., service centers). Recently, a connectivity constraint network Voronoi diagram (CCNVD) has been proposed to add capacity constraint to each partition while maintaining spatial continuity [127]. METIS [128] provides a set of scalable graph partitioning algorithms, which have shown high partition quality and computational efficiency.

2.3.4.2   Spatiotemporal Partitioning (clustering)

Spatiotemporal event partitioning (clustering): Most methods for 2-D spatial point clustering [119] can be easily generalized to 3-D spatiotemporal event data [129]. For example, ST-DBSCAN [130] is a spatiotemporal extension of the density-based spatial clustering method DBSCAN. ST-GRID [131] is another example that extends grid-based spatial clustering methods into 3-D grids.

Spatial time series partitioning (clustering): Spatial time series clustering aims to divide the space into regions such that the similarity between time series within the same region is maximized. Global partitioning methods such as K-means, K-medoids, and EM, as well as the hierarchical methods, can be applied.

Common (dis)similarity measures include Euclidean distance, Pearson's correlation, and dynamic time warping (DTW) distance. More details can be found in a recent survey [132]. However, due to the high dimensionality of spatial time series, density-based approaches and graph-based approaches are often not used. When computing similarities between spatial time series, a filter-and-refine approach [27] can be used to avoid redundant computation.

Trajectory partitioning: Trajectory partitioning approaches can be categorized by their objectives, namely trajectory grouping, flock pattern detection, and trajectory segmentation. Trajectory grouping aims to partition trajectories into groups according to their similarity. There are mainly two types of approaches, i.e., distance-based and frequency-based. The *density-based approaches* [133–135] first break trajectories into small segments and apply distance-based clustering algorithms similar to K-means or DBSCAN to connect dense areas of segments. The *frequency-based approach* [136] uses association rule mining [40] algorithms to identify subsections of trajectories which have high frequencies (also called high "support").

2.3.4.3 Spatial and Spatiotemporal Summarization

Data summarization aims to find compact representation of a dataset [137]. It is important for data compression as well as for making pattern analysis more convenient. Summarization can be done on classical data, spatial data, as well as spatiotemporal data.

Classical data summarization: Classical data can be summarized with aggregation statistics such as count, mean, and median. Many modern database systems provide query support for this operation, e.g., "Group by" operator in SQL.

Spatial data summarization: Spatial data summarization is more difficult than classical data summarization due to its non-numeric nature. For Euclidean space, the task can be done by first conducting spatial partitioning and then identifying representative spatial objects. For example, spatial data can be summarized with the centroids or medoids computed from K-means or K-medoids algorithms. For network space, especially for spatial network activities, summarization can be done by identifying several primary routes that cover those activities as much as possible. A K-Main Routes (KMR) algorithm [138] has been proposed to efficiently compute such routes to summarize spatial network activities. To reduce the computational cost, the KMR algorithm uses network Voronoi diagrams, divide and conquer, and pruning techniques.

Spatiotemporal data summarization: For spatial time series data, summarization can be done by removing spatial and temporal redundancy due to the effect of autocorrelation. A family of such algorithms has been used to summarize traffic data streams [139]. Similarly, the centroids from K-means can also be used to summarize spatial time series. For trajectory data, especially spatial network trajectories, summarization is more challenging due to the huge cost of similarity computation. A recent approach summarizes network trajectories into k-primary corridors

[140, 141]. The work proposes efficient algorithms to reduce the huge cost for network trajectory distance computation.

2.3.5   Spatial and Spatiotemporal Hotspot Detection

Problem definition: Given a set of spatial objects (e.g., points) in a study area, the problem of *spatial hotspot detection* aims to find regions where the number of objects is unexpectedly or anomalously high. Spatial hotspot detection is different from spatial partitioning or clustering, since spatial hotspots are a special kind of clusters whose intensity is "significantly" higher than the outside. *Spatiotemporal hotspots* can be seen as a generalization of spatial hotspots with a specified time window.

Challenges: Spatial and spatiotemporal hotspot detection is a challenging task since the location, size, and shape of a hotspot are unknown beforehand. In addition, the number of hotspots in a study area is often not known either. Moreover, "false" hotspots that aggregate events only by chance should often be avoided, since these false hotspots impede proper response by authorities (e.g., wasting police resources). Thus, it is often important to test the statistical significance of candidate spatial or spatiotemporal hotspots.

Statistical foundation: Spatial (or spatiotemporal) scan statistics [53, 142] (also discussed in Sect. 3.1) are used to detect statistically significant hotspots from spatial (or spatiotemporal) datasets. It uses a window (or cylinder) to scan the space (or space–time) for candidate hotspots and performs hypothesis testing. The null hypothesis states that the spatial (or spatiotemporal) points are completely spatially random (a homogeneous Poisson point process). The alternative hypothesis states that the points inside of the window (or cylinder) have higher intensity of points than outside. A test statistic called log likelihood ratio is computed for each candidate hotspot, and the candidate with the highest likelihood ratio can be further evaluated by its significance value (i.e., p-value).

Computational approaches: The following subsections summarize common spatial and spatiotemporal hotspot detection approaches by different input data types.

2.3.5.1   Spatial Hotspot from Spatial Point Pattern

Spatial partitioning approaches: Spatial point partitioning or clustering methods (Sect. 4.4.1) can be used to identify candidate hotspot patterns. After this, statistical tools may be used to evaluate the statistical significance of candidate patterns. Many of these methods have been implemented in CrimeStat, a software package for crime hotspot analysis [143].

Spatial scan statistics based approaches: These approaches use a window with varying sizes to scan the 2-D plane and identifies the candidate window with the highest likelihood ratio. Statistical significance (p-value) is computed for this candidate based on Monte Carlo simulation. Scanning windows with different shapes, including circular, elliptical, as well as ring-shaped, have been proposed together with efficient computational pruning strategies [142, 144–146]. SaTScan [142] is a popular spatial scan statistics tool in epidemiology to analyze circular or elliptical hotspots.

Kernel Density Estimation: Kernel density estimation (KDE) [147] identifies spatial hotspots via a density map of point events. It first creates a grid over the study area and uses a kernel function with a user-defined radius (bandwidth) on each point to estimate the density of points on centers of grid cells. A subset of grid cells with high density are returned as spatial hotspots.

2.3.5.2 Spatial Hotspot from Areal Model

Local Indicators of Spatial Association: Local indicators of spatial association (LISA) [148, 149] is a set of local spatial autocorrelation statistics, including local Moran's I, Geary's C, or Ord Gi and Gi* functions. It differs from global spatial autocorrelation in that the statistics are computed within the neighborhood of a location. For example, a high local Moran's I indicates that values of the current location as well as its neighbors are both extremely high (or low) compared to values at other locations, and thus, the neighborhood is a spatial hotspot (or "cold spot").

2.3.5.3 Spatiotemporal Hotspot Detection

Hot routes from spatial network trajectories: Hot routes detection from spatial network trajectories aims to detect network paths with high density [133] or frequency of trajectories [136]. Other approaches include organizing police patrol routes [150], main streets [151], and clumping [152].

Spatiotemporal Scan Statistics based approaches: Two types of spatiotemporal hotspots can be detected by spatiotemporal scan statistics: "persistent" spatiotemporal hotspots and "emerging" spatiotemporal hotspots. A "persistent" spatiotemporal hotspot is a region where the rate of increase in observations is a high and almost constant value over time. Thus, approaches to detect a persistent spatiotemporal hotspot involves counting observations in each time interval [142]. An "emerging" spatiotemporal hotspot is a region where the rate of observations monotonically increases over time [145, 153]. This kind of spatiotemporal hotspot occurs when an outbreak emerges causing a sudden increase in the number observations. Tools for the detection of emerging spatiotemporal hotspots use spatial scan statistics to identify changes in expectation over time [154].

2.3.6 Spatiotemporal Change

2.3.6.1 What Are Spatiotemporal Changes and Change Footprints

Although the single term "change" is used to name the spatiotemporal change foot-print patterns in different applications, the underlying phenomena may differ signif-icantly. This section briefly summarizes the main ways a change may be defined and detected in spatiotemporal data [10].

Change in Statistical Parameter: In this case, the data is assumed to follow a cer-tain distribution and the change is defined as a shift in this statistical distribution. For example, in statistical quality control, a change in the mean or variance of the sensor readings is used to detect a fault.

Change in Actual Value: Here, change is modeled as the difference between a data value and its spatial or temporal neighborhood. For example, in a one-dimensional continuous function, the magnitude of change can be characterized by the derivative function, while on a two-dimensional surface, it can be characterized by the gradient magnitude.

Change in Models Fitted to Data: This type of change is identified when a number of models are fitted to the data and one or more of the models exhibits a change (e.g., a discontinuity between consecutive linear functions) [155].

2.3.6.2 Common Approaches

This section follows the taxonomy of spatiotemporal change footprint patterns as proposed in [10]. In this taxonomy, spatiotemporal change footprints are classified along two dimensions: temporal and spatial. Temporal footprints are classified into four categories: single snapshot, set of snapshots, point in a long series, and interval in a long series. Single snapshot refers to a purely spatial change that does not have a temporal context. A set of snapshots indicate a change between two or more snapshots of the same spatial field, e.g., satellite images of the same region.

Spatial footprints can be classified as raster footprints or vector footprints. Vector footprints are further classified into four categories: point(s), line(s), polygon(s), and network footprint patterns. Raster footprints are classified based on the scale of the pattern, namely local, focal, or zonal patterns. This classification describes the scale of the change operation of a given phenomenon in the spatial raster field [156]. Local patterns are patterns in which change at a given location depends only on attributes at this location. Focal patterns are patterns in which change in a location depends on attributes in that location and its assumed neighborhood. Zonal patterns define change using an aggregation of location values in a region.

Spatiotemporal Change Patterns with Raster-based Spatial Footprint: This includes patterns of *spatial changes between snapshots*. In remote sensing, detecting changes between satellite images can help identify land cover change due to human activity, natural disasters, or climate change [157–159]. Given two geographically aligned raster images, this problem aims to find a collection of pixels that have significant changes between the two images [160]. This pattern is classified as a local change between snapshots since the change at a given pixel is assumed to be independent of changes at other pixels. Alternative definitions have assumed that a change at a pixel also depends on its neighborhoods [161]. For example, the pixel values in each block may be assumed to follow a Gaussian distribution [162]. We refer to this type of change footprint pattern as a focal spatial change between snapshots. Researchers in remote sensing and image processing have also tried to apply image change detection to objects instead of pixels [163–165], yielding zonal spatial change patterns between snapshots.

A well-known technique for detecting a local change footprint is simple differencing. The technique starts by calculating the differences between the corresponding pixels intensities in the two images. A change at a pixel is flagged if the difference at the pixel exceeds a certain threshold. Alternative approaches have also been proposed to discover focal change footprints between images. For example, the block-based density ratio test detects change based on a group of pixels, known as a block [166, 167]. Object-based approaches in remote sensing [165, 168, 169] employ image segmentation techniques to partition temporal snapshots of images into homogeneous objects [170] and then classify object pairs in the two temporal snapshots of images into no change or change classes.

Spatiotemporal Change Patterns with Vector-based Spatial Footprint: This includes the *Spatiotemporal Volume Change Footprint* pattern. This pattern represents a change process occurring in a spatial region (a polygon) during a time interval. For example, an outbreak event of a disease can be defined as an increase in disease reports in a certain region during a certain time window up to the current time. Change patterns known to have an spatiotemporal volume footprint include the spatiotemporal scan statistics [171, 172], a generalization of the spatial scan statistic, and emerging spatiotemporal clusters defined by [154].

2.4 Research Trend and Future Research Needs

Most current research in spatial and spatiotemporal data science uses Euclidean space, which often assumes isotropic property and symmetric neighborhoods. However, in many real-world applications, the underlying space is network space, such as river networks and road networks [138, 173, 174]. One of the main challenges in spatial and spatiotemporal network data science is to account for the network structure in the dataset. For example, in anomaly detection, spatial techniques do not consider the spatial network structure of the dataset; that is, they may not be able to model graph properties such as one ways, connectivity, and left-turns. The network structure often violates the isotropic property and symmetry of neighborhoods and instead requires asymmetric neighborhood and directionality of neighborhood relationship (e.g., network flow direction).

Recently, some cutting edge research has been conducted in the spatial network statistics and data science [57]. For example, several spatial network statistical methods have been developed, e.g., network K-function and network spatial autocorrelation. Several spatial analysis methods have also been generalized to the network space, such as network point cluster analysis and clumping method, network point density estimation, network spatial interpolation (Kriging), as well as network Huff model. Due to the nature of spatial network space as distinct from Euclidean space, these statistics and analysis often rely on advanced spatial network computational techniques [57].

We believe more spatial and spatiotemporal big data science research is still needed in the network space. First, though several spatial statistics and big data science techniques have been generalized to the network space, few spatiotemporal network statistics and big data science have been developed, and the vast majority of research is still in the Euclidean space. Future research is needed to develop more spatial network statistics, such as spatial network scan statistics, spatial network random field model, as well as spatiotemporal autoregressive models for networks. Furthermore, phenomena observed on spatiotemporal networks need to be interpreted in an appropriate frame of reference to prevent a mismatch between the nature of the observed phenomena and the mining algorithm. For instance, moving objects on a spatiotemporal network need to be studied from a traveler's perspective, i.e., the Lagrangian frame of reference [175–178] instead of a snapshot view. This is because a traveler moving along a chosen path in a spatiotemporal network would experience a road segment (and its properties such as fuel efficiency and travel time) for the time at which he/she arrives at that segment, which may be distinct from the original departure time at the start of the journey. These unique requirements (non-isotropy and Lagrangian reference frame) call for novel spatiotemporal statistical foundations [173] as well as new computational approaches for spatiotemporal network big data science.

Another future research need is to develop spatiotemporal graph big data platforms, motivated by the upcoming rich spatiotemporal network data collected from vehicles. Modern vehicles have rich instrumentation to measure hundreds of attributes at high frequency and are generating big data (Exabyte [179]). This vehicle measurement big data consists of a collection of trips on a transportation graph such as a road map annotated with several measurements of engine subsystems. Collecting and analyzing such big data during real-world driving conditions can aid in understanding the underlying factors which govern real-world fuel inefficiencies or high greenhouse gas emissions [180]. Current relevant big data platforms for spatial and spatiotemporal big data science include ESRI GIS Tools for Hadoop [181, 182] and Hadoop GIS [183]. These provide distributed systems for geometric data (e.g., lines, points, and polygons) including geometric indexing and partitioning methods such as R-tree, R+-tree, or Quad tree. Recently, SpatialHadoop has been developed [184]. SpatialHadoop embeds geometric notions in language, visualization, storage, MapReduce, and operations layers. However, spatiotemporal graphs violate the core assumptions of current spatial big data platforms that the geometric concepts are adequate for conveniently representing spatiotemporal graph analytics operations and for partition data for load-balancing. Spatiotempo-ral graphs also violate core assumptions underlying graph analytics software (e.g., Giraph [185], GraphLab [186], and Pregel [187]) that traditional location-unaware graphs are adequate for conveniently representing STG analytics operations and for partition data for load-balancing. Therefore, novel spatiotemporal graph big data platforms is needed. Several challenges should be addressed; e.g., spatiotemporal graph big data requires novel distributed file system (DFS) to partition the graph, and a novel programming model is still needed to support abstract data types and fundamental spatiotemporal graphs operations, etc.

2.5   Summary

This chapter provides an overview of current research in the field of spatial and spatiotemporal (SST) big data science from a computational perspective. SST big data science has broad application domains including ecology and environmental management, public safety, transportation, earth science, epidemiology, and clima-tology. However, the complexity of SST data and intrinsic SST relationships limits the usefulness of conventional big data science techniques. We provide a taxonomy of different SST data types and underlying statistics. We also review common SST big data science techniques organized by major output pattern families: SST outlier, cou-pling and tele-coupling, prediction, partitioning and summarization, hotspots, and change patterns. Finally, we discuss the recent research trends and future research needs.

References

1. S. Shekhar, Z. Jiang, R.Y. Ali, E. Eftelioglu, X. Tang, V.M.V. Gunturi, X. Zhou, Spatiotemporal data mining: a computational perspective. ISPRS Int. J. Geo-Inf. **4**(4), 2306 (2015)
2. K. Koperski, J. Adhikary, J. Han, Spatial data mining: progress and challenges survey paper, in *Proceedings of ACM SIGMOD Workshop on Research Issues on Data Mining and Knowledge Discovery, Montreal, Canada* (Citeseer, 1996), pp. 1–10
3. M. Ester, H.-P. Kriegel, J. Sander, Spatial data mining: a database approach, in *Proceedings of Fifth Symposium on Rules in Geographic Information Databases* (1997)
4. S. Shekhar, M.R. Evans, J.M. Kang, P. Mohan, Identifying patterns in spatial information: a survey of methods. Wiley Interdis. Rev. Data Min. Knowl. Disc. **1**(3), 193–214 (2011)
5. H.J. Miller, J. Han, *Geographic Data Mining and Knowledge Discovery* (Taylor & Francis Inc., Bristol, 2001)
6. H.J. Miller, J. Han, in *Geographic Data Mining and Knowledge Discovery* (CRC Press, 2009)
7. S. Shekhar, P. Zhang, Y. Huang, R.R. Vatsavai, Trends in spatial data mining, in *Data Mining: Next Generation Challenges and Future Directions* (2003), pp. 357–380
8. S. Kisilevich, F. Mansmann, M. Nanni, S. Rinzivillo, in *Spatio-Temporal Clustering* (Springer, Berlin, 2010)
9. C.C. Aggarwal, in *Outlier Analysis* (Springer Science & Business Media, 2013)
10. X. Zhou, S. Shekhar, R.Y. Ali, Spatiotemporal change footprint pattern discovery: an inter-disciplinary survey. Wiley Interdis. Rev. Data Min. Knowl. Disc. **4**(1), 1–23 (2014)

11. A. Karpatne, Z. Jiang, R.R. Vatsavai, S. Shekhar, V. Kumar, Monitoring land-cover changes: A machine-learning perspective. IEEE Geosci. Rem. Sens. Mag. **4**(2), 8–21 (2016)
12. S. Shekhar, S. Chawla, in *Spatial Databases: A Tour* (Prentice Hall, Englewood-Cliffs, 2003)
13. M. Worboys, M. Duckham, in *GIS: A Computing Perspective*, 2nd edn. (CRC, 2004). ISBN: 978-0415283755
14. Z. Li, J. Chen, E. Baltsavias, in *Advances in Photogrammetry, Remote Sensing and Spatial Information Sciences: 2008 ISPRS Congress Book*, vol 7 (CRC Press, 2008)
15. M. Yuan, Temporal gis and spatio-temporal modeling, in *Proceedings of Third International Conference Workshop on Integrating GIS and Environment Modeling, Santa Fe, NM* (1996)
16. J.F. Allen, Towards a general theory of action and time. Artif. Intell. **23**(2), 123–154 (1984)
17. B. George, S. Kim, S. Shekhar, Spatio-temporal network databases and routing algorithms: a summary of results, in *Proceedings of International Symposium on Spatial and Temporal Databases (SSTD'07)* (Boston, 2007)
18. B. George, S. Shekhar, Time aggregated graphs: a model for spatio-temporal network, in *Proceedings of the Workshops (CoMoGIS) at the 25th International Conference on Conceptual Modeling (ER2006)* (Tucson, AZ, USA, 2006)
19. A.E. Gelfand, P. Diggle, P. Guttorp, M. Fuentes, in *Handbook of Spatial Statistics* (CRC Press, 2010)
20. C.E. Campelo, B. Bennett, in *Representing and Reasoning About Changing Spatial Extensions of Geographic Features* (Springer, Berlin, 2013)
21. P. Tan, M. Steinbach, V. Kumar, et al., in *Introduction to Data Mining* (Pearson Addison Wesley Boston, 2006)
22. P. Bolstad, in *GIS Fundamentals: A First Text on GIS* (Eider Press, 2002)
23. A.R. Ganguly, K. Steinhaeuser, Data mining for climate change and impacts, in *ICDM Workshops* (2008), pp. 385–394
24. M. Erwig, M. Schneider, F. Hagen, Spatio-temporal predicates. IEEE Trans. Knowl. Data Eng. **14**, 881–901 (2002)
25. J. Chen, R. Wang, L. Liu, J. Song, Clustering of trajectories based on hausdorff distance, in *2011 International Conference on Electronics, Communications and Control (ICECC)* (IEEE, 2011), pp. 1940–1944
26. Z. Zhang, K. Huang, T. Tan, Comparison of similarity measures for trajectory clustering in outdoor surveillance scenes, in *18th International Conference on Pattern Recognition, 2006. ICPR 2006*, vol. 3 (IEEE, 2006), pp. 1135–1138
27. P. Zhang, Y. Huang, S. Shekhar, V. Kumar, Correlation analysis of spatial time series datasets: a filter-and-refine approach, in *Advances in Knowledge Discovery and Data Mining* (Springer, Berlin, 2003), pp. 532–544
28. J. Kawale, S. Chatterjee, D. Ormsby, K. Steinhaeuser, S. Liess, V. Kumar, Testing the significance of spatio-temporal teleconnection patterns, in *KDD* (2012), pp. 642–650
29. M. Celik, S. Shekhar, J.P. Rogers, J.A. Shine, J.S. Yoo, Mixed-drove spatio-temporal co-occurrence pattern mining: a summary of results, in *ICDM '06: Proceedings of the Sixth International Conference on Data Mining* (IEEE Computer Society, Washington, DC, USA, 2006), pp. 119–128
30. K.G. Pillai, R.A. Angryk, B. Aydin, A filter-and-refine approach to mine spatiotemporal co-occurrences, in *SIGSPATIAL/GIS* (2013), pp. 104–113
31. P. Mohan, S. Shekhar, J.A. Shine, J.P. Rogers, Cascading spatio-temporal pattern discovery. IEEE Trans. Knowl. Data Eng. **24**(11), 1977–1992 (2012)
32. P. Mohan, S. Shekhar, J.A. Shine, J.P. Rogers, Cascading spatio-temporal pattern discovery: a summary of results in *SDM* (2010), pp. 327–338
33. Y. Huang, L. Zhang, P. Zhang, A framework for mining sequential patterns from spatio-temporal event data sets. IEEE Trans. Knowl. Data Eng. **20**(4), 433–448 (2008)
34. Y. Huang, L. Zhang, P. Zhang, Finding sequential patterns from a massive number of spatio-temporal events, in *SDM* (2006), pp. 634–638
35. J. Mennis, R. Viger, C.D. Tomlin, Cubic map algebra functions for spatio-temporal analysis. Cartography Geogr. Inf. Sci. **32**(1), 17–32 (2005)

36. D.G. Brown, R. Riolo, D.T. Robinson, M. North, W. Rand, Spatial process and data models: toward integration of agent-based models and gis. J. Geogr. Syst. **7**(1), 25–47 (2005)
37. J. Quinlan, in *C4.5: Programs for Machine Learning* (Morgan Kaufmann Publishers, 1993)
38. V. Varnett, T. Lewis, in *Outliers in Statistical Data* (Wiley, New York, 1994)
39. T. Agarwal, R. Imielinski, A. Swami, Mining association rules between sets of items in large databases, in *Proceedings of the ACM SIGMOD Conference on Management of Data* (Washington, D.C., 1993)
40. R. Agrawal, R. Srikant, Fast algorithms for mining association rules, in *Proceedings of Very Large Databases* (1994)
41. A. Jain, R. Dubes, in *Algorithms for Clustering Data* (Prentice Hall, 1988)
42. S. Banerjee, B. Carlin, A. Gelfand, in *Hierarchical Modeling and Analysis for Spatial Data* (Chapman & Hall, 2004)
43. O. Schabenberger, C. Gotway, in *Statistical Methods for Spatial Data Analysis* (Chapman and Hall, 2005)
44. N.A.C. Cressie, in *Statistics for Spatial Data* (Wiley, New York, 1993)
45. S. Banerjee, B.P. Carlin, A.E. Gelfrand, in *Hierarchical Modeling and Analysis for Spatial Data* (CRC Press, 2003)
46. A. Fotheringham, C. Brunsdon, M. Charlton, in *Geographically Weighted Regression: The Analysis of Spatially Varying Relationships* (Wiley, New York, 2002)
47. C.E. Warrender, M.F. Augusteijn, Fusion of image classifications using Bayesian techniques with Markov rand fields. Int. J. Remote Sens. **20**(10), 1987–2002 (1999)
48. N. Cressie, *Statistics for Spatial Data*, Revised edn. (Wiley, New York, 1993)
49. L. Anselin, Local indicators of spatial association-lisa. Geograp. Anal. **27**(2), 93–155 (1995)
50. S. Openshaw, in *The Modifiable Areal Unit Problem*, (OCLC, 1983), ISBN: 0860941345
51. B.D. Ripley, Modelling spatial patterns, in*Journal of the Royal Statistical Society. Series B (Methodological)* (1977), pp. 172–212
52. E. Marcon, F. Puech, et al., Generalizing Ripley's k function to inhomogeneous populations. Technical report (Mimeo, 2003)
53. M. Kulldorff, A spatial scan statistic. Commun. Stat. Theor. Methods **26**(6), 1481–1496 (1997)
54. S.N. Chiu, D. Stoyan, W.S. Kendall, J. Mecke, in *Stochastic Geometry and Its Applications* (Wiley, 2013)
55. X. Guyon, in *Random Fields on a Network: Modeling, Statistics, and Applications* (Springer Science & Business Media, 1995)
56. A. Okabe, H. Yomono, M. Kitamura, Statistical analysis of the distribution of points on a network. Geograph. Anal. **27**, 152–175 (1995)
57. A. Okabe, K. Sugihara, in *Spatial Analysis Along Networks: Statistical and Computational Methods* (Wiley, New York, 2012)
58. A. Okabe, K. Okunuki, S. Shiode, The sanet toolbox: new methods for network spatial analysis. Trans. GIS **10**(4), 535–550 (2006)
59. N. Cressie, C.K. Wikle, in *Statistics for Spatio-Temporal Data* (Wiley, New York, 2011)
60. R.H. Shumway, D.S. Stoffer, in *Time Series Analysis and Its Applications: With R Examples* (Springer Science & Business Media, 2010)
61. P.C. Kyriakidis, A.G. Journel, Geostatistical space-time models: a review. Math. Geol. **31**(6), 651–684 (1999)
62. N.A.C. Cressie, in *Statistics for Spatial Data* (Wiley, New York, 1993), ISBN: 978-0471002550
63. V. Barnett, T. Lewis, in *Outliers in Statistical Data*, 3rd edn. (Wiley, New York, 1994)
64. V. Chandola, A. Banerjee, V. Kumar, Anomaly detection: a survey. ACM Comput. Surv. **41**(3), 15:1–15:58 (2009)
65. S. Shekhar, C. Lu, P. Zhang, A unified approach to detecting spatial outliers. GeoInformatica **7**(2), 139–166 (2003)
66. J. Haslett, R. Bradley, P. Craig, A. Unwin, G. Wills, Dynamic graphics for exploring spatial data with application to locating global and local anomalies, in *American Statistician* (1991), pp. 234–242

67. A. Luc, Exploratory spatial data analysis and geographic information systems, in *New Tools for Spatial Analysis*, ed. by M. Painho (1994), pp. 45–54
68. D. Chen, C.-T. Lu, Y. Kou, F. Chen, On detecting spatial outliers. GeoInformatica **12**(4), 455–475 (2008)
69. C.-T. Lu, D. Chen, Y. Kou, Detecting spatial outliers with multiple attributes, in *ICTAI '03: Proceedings of the 15th IEEE International Conference on Tools with Artificial Intelligence* (IEEE Computer Society, Washington, DC, USA, 2003), p. 122
70. Y. Kou, C.-T. Lu, D. Chen, Spatial weighted outlier detection, in *SDM* (2006), pp. 614–618
71. X. Liu, F. Chen, C.-T. Lu, On detecting spatial categorical outliers. GeoInformatica **18**(3), 501–536 (2014)
72. E. Schubert, A. Zimek, H.-P. Kriegel, Local outlier detection reconsidered: a generalized view on locality with applications to spatial, video, and network outlier detection. Data Min. Knowl. Discov. **28**(1), 190–237 (2014)
73. M. Wu, C. Jermaine, S. Ranka, X. Song, J. Gums, A model-agnostic framework for fast spatial anomaly detection. TKDD **4**(4), 20 (2010)
74. A.M. Sainju, Z. Jiang. Grid-based co-location mining algorithms on GPU for big spatial event data: a summary of results, in *Proceedings of International Symposium on Spatial and Temporal Databases (SSTD)*, (2017 to appear)
75. J.M. Kang, S. Shekhar, C. Wennen, P. Novak, Discovering flow anomalies: a SWEET approach, in *International Conference on Data Mining* (2008)
76. Y. Huang, S. Shekhar, H. Xiong, Discovering co-location patterns from spatial datasets: a general approach. IEEE Trans. Knowl. Data Eng. (TKDE) **16**(12), 1472–1485 (2004)
77. M. Celik, S. Shekhar, J.P. Rogers, J.A. Shine, Mixed-drove spatiotemporal co-occurrence pattern mining. IEEE Trans. Knowl. Data Eng. **20**(10), 1322–1335 (2008)
78. Y. Chou, in *Exploring Spatial Analysis in Geographic Information System* (Onward Press, 1997)
79. K. Koperski, J. Han, Discovery of Spatial Association Rules in Geographic Information Databases, in *Proceedings of Fourth International Symposium on Large Spatial Databases* (Maine, 1995), pp. 47–66
80. Y. Morimoto, Mining frequent neighboring class sets in spatial databases, in *ACM SIGKDD International Conference on Knowledge Discovery and Data Mining* (2001)
81. H. Xiong, S. Shekhar, Y. Huang, V. Kumar, X. Ma, J.S. Yoo, A framework for discovering co-location patterns in data sets with extended spatial objects, in *SDM* (2004), pp. 78–89
82. Y. Huang, J. Pei, H. Xiong, Mining co-location patterns with rare events from spatial data sets. GeoInformatica **10**(3), 239–260 (2006)
83. S. Wang, Y. Huang, X.S. Wang, Regional co-locations of arbitrary shapes, in *SSTD* (2013), pp. 19–37
84. W. Ding, C.F. Eick, X. Yuan, J. Wang, J.-P. Nicot, A framework for regional association rule mining and scoping in spatial datasets. GeoInformatica **15**(1), 1–28 (2011)
85. P. Mohan, S. Shekhar, J.A. Shine, J.P. Rogers, Z. Jiang, N. Wayant, A neighborhood graph based approach to regional co-location pattern discovery: a summary of results, in *GIS* (2011), pp. 122–132
86. S. Barua, J. Sander, Mining statistically significant co-location and segregation patterns. IEEE Trans. Knowl. Data Eng. **26**(5), 1185–1199 (2014)
87. J.S. Yoo, S. Shekhar, A joinless approach for mining spatial colocation patterns. IEEE Trans. Knowl. Data Eng. (TKDE) **18**(10), 1323–1337 (2006)
88. H. Cao, N. Mamoulis, D.W. Cheung, Discovery of collocation episodes in spatiotemporal data, in *ICDM* (2006), pp. 823–827
89. H. Cao, N. Mamoulis, D.W. Cheung, Mining frequent spatio-temporal sequential patterns, in *ICDM* (2005), pp. 82–89
90. F. Verhein, Mining complex spatio-temporal sequence patterns, in *SDM* (2009), pp. 605–616
91. L.A. Tang, Y. Zheng, J. Yuan, J. Han, A. Leung, W.-C. Peng, T.F.L. Porta, A framework of traveling companion discovery on trajectory data streams. ACM TIST **5**(1), 3 (2013)

92. W.R. Tobler, A computer movie simulating urban growth in the detroit region. Econ. Geograph. **46**, 234–240 (1970)

93. I. Vainer, S. Kraus, G. Kaminka, H. Slovin, Scalable classification in large scale spatiotemporal domains applied to voltage-sensitive dye imaging, in *Ninth IEEE International Conference on Data Mining, 2009. ICDM '09* (2009), pp. 543–551

94. M. Ceci, A. Appice, D. Malerba, Spatial associative classification at different levels of granularity: a probabilistic approach, in *PKDD* (2004), pp. 99–111

95. W. Ding, T.F. Stepinski, J. Salazar, Discovery of geospatial discriminating patterns from remote sensing datasets, in *SDM* (SIAM, 2009), pp. 425–436

96. R. Frank, M. Ester, A.J. Knobbe, A multi-relational approach to spatial classification, in *KDD* (2009), pp. 309–318

97. M.D. Twa, S. Parthasarathy, T.W. Raasch, M. Bullimore, Decision tree classification of spatial data patterns from videokeratography using zernicke polynomials, in *SDM* (2003), pp. 3–12

98. J. Li, A.D. Heap, A review of comparative studies of spatial interpolation methods in environmental sciences: performance and impact factors. Ecol. Inf. **6**(3), 228–241 (2011)

99. S. Bhattacharjee, P. Mitra, S.K. Ghosh, Spatial interpolation to predict missing attributes in GIS using semantic kriging. IEEE Trans. Geosci. Remote Sens. **52**(8), 4771–4780 (2014)

100. A.K. Bhowmik, P. Cabral, Statistical evaluation of spatial interpolation methods for small-sampled region: a case study of temperature change phenomenon in bangladesh, in *Computational Science and Its Applications-ICCSA 2011* (Springer, Berlin, 2011), pp. 44–59

101. S. Li, in *A Markov Random Field Modeling* (Computer Vision Publisher, Springer, 1995)

102. S. Shekhar, P.R. Schrater, R.R. Vatsavai, W. Wu, S. Chawla, Spatial Contextual Classification and Prediction Models for Mining Geospatial Data. IEEE Trans. Multimedia **4**(2), 174–188 (2002)

103. C.-H. Lee, R. Greiner, O.R. Zaïane, Efficient spatial classification using decoupled conditional random fields, in *PKDD* (2006), pp. 272–283

104. L. Anselin, *Spatial Econometrics: Methods and Models* (Kluwer, Dordrecht, 1988)

105. S. Chawla, S. Shekhar, W.-L. Wu, U. Ozesmi, Modeling spatial dependencies for mining geospatial data. ACM SIGMOD Workshop Res. Issues Data Min. Knowl. Disc. **70–77**, 2000 (2000)

106. S. Chawla, S. Shekhar, W. Wu, U. Ozesmi, Modeling spatial dependencies for mining geospatial data, in *1st SIAM International Conference on Data Mining* (2001)

107. A. Liu, G. Jun, J. Ghosh, Spatially cost-sensitive active learning, in *SDM* (SIAM, 2009), pp. 814–825

108. K. Subbian, A. Banerjee, Climate multi-model regression using spatial smoothing, in *SDM* (2013), pp. 324–332

109. A. McGovern, N. Troutman, R.A. Brown, J.K. Williams, J. Abernethy, Enhanced spatiotemporal relational probability trees and forests. Data Min. Knowl. Discov. **26**(2), 398–433 (2013)

110. J.-G. Lee, J. Han, X. Li, H. Cheng, Mining discriminative patterns for classifying trajectories on road networks. IEEE Trans. Knowl. Data Eng. **23**(5), 713–726 (2011)

111. A. Noulas, S. Scellato, N. Lathia, C. Mascolo, Mining user mobility features for next place prediction in location-based services, in *ICDM* (2012), pp. 1038–1043

112. J.J.-C. Ying, W.-C. Lee, V.S. Tseng, Mining geographic-temporal-semantic patterns in trajectories for location prediction. ACM TIST **5**(1), 2 (2013)

113. H. Cheng, J. Ye, Z. Zhu, What's your next move: User activity prediction in location-based social networks, in *SDM* (2013), pp. 171–179

114. J.-D. Zhang, C.-Y. Chow, iGSLR: personalized geo-social location recommendation: a kernel density estimation approach, in *SIGSPATIAL/GIS* (2013), pp. 324–333

115. B. Liu, Y. Fu, Z. Yao, H. Xiong, Learning geographical preferences for point-of-interest recommendation, in *KDD* (2013), pp. 1043–1051

116. Y. Zheng, X. Xie, Learning travel recommendations from user-generated GPS traces. ACM TIST **2**(1), 2 (2011)

117. H. Wang, M. Terrovitis, N. Mamoulis, Location recommendation in location-based social networks using user check-in data, in *SIGSPATIAL/GIS* (2013), pp. 364–373

118. J. Bao, Y. Zheng, M.F. Mokbel, Location-based and preference-aware recommendation using sparse geo-social networking data, in *SIGSPATIAL/GIS* (2012), pp. 199–208
119. J. Han, M. Kamber, A.K.H. Tung, Spatial Clustering Methods in Data Mining: A Survey, in *Geographic Data Mining and Knowledge Discovery* (Taylor and Francis, 2001)
120. G. Karypis, E.-H. Han, V. Kumar, Chameleon: hierarchical clustering using dynamic modeling. IEEE Comput. **32**(8), 68–75 (1999)
121. M. Ester, H.-P. Kriegel, J. Sander, X. Xu, A density-based algorithm for discovering clusters in large spatial databases with noise. KDD **96**, 226–231 (1996)
122. R.A. Jarvis, E.A. Patrick, Clustering using a similarity measure based on shared near neighbors. IEEE Trans. Comput. **100**(11), 1025–1034 (1973)
123. M. Worboys, in *GIS: A Computing Perspective* (Taylor and Francis, 1995)
124. D. Joshi, A. Samal, L.-K. Soh, A dissimilarity function for clustering geospatial polygons, in *Proceedings of the 17th ACM SIGSPATIAL International Conference on Advances in Geographic Information Systems* (ACM, 2009), pp. 384–387
125. S. Wang, C.F. Eick, A polygon-based clustering and analysis framework for mining spatial datasets. GeoInformatica **18**(3), 569–594 (2014)
126. R.M. Haralick, L.G. Shapiro, Image segmentation techniques, in *1985 Technical Symposium East* (International Society for Optics and Photonics, 1985), pp. 2–9
127. K. Yang, A.H. Shekhar, D. Oliver, S. Shekhar, Capacity-constrained network-voronoi diagram. IEEE Trans. Knowl. Data Eng. **27**(11), 2919–2932 (2015)
128. G. Karypis, Multi-constraint mesh partitioning for contact/impact computations, in *Proceedings of the 2003 ACM/IEEE Conference on Supercomputing* (ACM, 2003), p. 56
129. D. Joshi, A. Samal, L.-K. Soh, Spatio-temporal polygonal clustering with space and time as first-class citizens. GeoInformatica **17**(2), 387–412 (2013)
130. D. Birant, A. Kut, St-dbscan: an algorithm for clustering spatial-temporal data. Data Knowl. Eng. **60**(1), 208–221 (2007)
131. M. Wang, A. Wang, A. Li, Mining spatial-temporal clusters from geo-databases, in *Advanced Data Mining and Applications* (Springer, Berlin, 2006), pp. 263–270
132. T.W. Liao, Clustering of time series data-a survey. Pattern Recogn. **38**(11), 1857–1874 (2005)
133. J.-G. Lee, J. Han, K.-Y. Whang, Trajectory clustering: a partition-and-group framework, in *Proceedings of the 2007 ACM SIGMOD International Conference on Management of Data* (ACM, 2007), pp. 593–604
134. Z. Zhang, Y. Yang, A.K. Tung, D. Papadias, Continuous k-means monitoring over moving objects. IEEE Trans. Knowl. Data Eng. **20**(9), 1205–1216 (2008)
135. C.S. Jensen, D. Lin, B.C. Ooi, Continuous clustering of moving objects. IEEE Trans. Knowl. Data Eng. **19**(9), 1161–1174 (2007)
136. A.J. Lee, Y.-A. Chen, W.-C. Ip, Mining frequent trajectory patterns in spatial-temporal databases. Inf. Sci. **179**(13), 2218–2231 (2009)
137. V. Chandola, V. Kumar, Summarization-compressing data into an informative representation. Knowl. Inf. Syst. **12**(3), 355–378 (2007)
138. D. Oliver, S. Shekhar, J.M. Kang, R. Laubscher, V. Carlan, A. Bannur, A k-main routes approach to spatial network activity summarization. IEEE Trans. Knowl. Data Eng. **26**(6), 1464–1478 (2014)
139. B. Pan, U. Demiryurek, F. Banaei-Kashani, C. Shahabi, Spatiotemporal summarization of traffic data streams, in *Proceedings of the ACM SIGSPATIAL International Workshop on GeoStreaming* (ACM, 2010), pp. 4–10
140. M.R. Evans, D. Oliver, S. Shekhar, F. Harvey, Summarizing trajectories into k-primary corridors: a summary of results, in *Proceedings of the 20th International Conference on Advances in Geographic Information Systems* (ACM, 2012), pp. 454–457
141. Z. Jiang, M. Evans, D. Oliver, S. Shekhar, Identifying K primary corridors from urban bicycle GPS trajectories on a road network. Inf. Syst. (2015) (to appear)
142. M. Kulldorff, Satscan user guide for version. **9**, 4–107 (2011)
143. N. Levine, in *CrimeStat 3.0: A Spatial Statistics Program for the Analysis of Crime Incident Locations* (Ned Levine & Associatiates: Houston, TX/National Institute of Justice: Washington, DC, 2004)

144. E. Eftelioglu, S. Shekhar, D. Oliver, X. Zhou, M.R. Evans, Y. Xie, J.M. Kang, R. Laubscher, C. Farah, Ring-shaped hotspot detection: a summary of results, in *2014 IEEE International Conference on Data Mining, ICDM 2014, Shenzhen, China, December 14–17, 2014* (2014), pp. 815–820

145. T. Tango, K. Takahashi, K. Kohriyama, A space-time scan statistic for detecting emerging outbreaks. Biometrics **67**(1), 106–115 (2011)

146. D.B. Neill, A.W. Moore, A fast multi-resolution method for detection of significant spatial disease clusters, in *Advances in Neural Information Processing Systems* (2003)

147. J. Ratcliffe, Crime mapping: spatial and temporal challenges, in *Handbook of Quantitative Criminology* (Springer, Berlin, 2010), pp. 5–24

148. A. Luc, Local indicators of spatial association: LISA. Geograph. Anal. **27**(2), 93–115 (1995)

149. N. Chaikaew, N.K. Tripathi, M. Souris, International journal of health geographics. Int. J. Health Geograph. **8**, 36 (2009)

150. S.S. Chawathe, Organizing hot-spot police patrol routes, in *Intelligence and Security Informatics, 2007 IEEE* (IEEE, 2007), pp. 79–86

151. M. Celik, S. Shekhar, B. George, J.P. Rogers, J.A. Shine, Discovering and quantifying mean streets: a summary of results. Technical Report 025 (University of Minnesota, 07 2007)

152. S. Shiode, A. Okabe, Network variable clumping method for analyzing point patterns on a network, in *Unpublished Paper Presented at the Annual Meeting of the Associations of American Geographers* (Philadelphia, Pennsylvania, 2004)

153. W. Chang, D. Zeng, H. Chen, Prospective spatio-temporal data analysis for security informatics, in *Intelligent Transportation Systems, 2005. Proceedings. 2005 IEEE* (IEEE, 2005), pp. 1120–1124

154. D. Neill, A. Moore, M. Sabhnani, K. Daniel, Detection of emerging space-time clusters, in *Proceedings of the eleventh ACM SIGKDD International Conference on Knowledge Discovery in Data Mining* (ACM, 2005), pp. 218–227

155. V. Chandola, D. Hui, L. Gu, B. Bhaduri, R. Vatsavai, Using time series segmentation for deriving vegetation phenology indices from MODIS NDVI data, in *IEEE International Conference on Data Mining Workshops* (Sydney, Australia, 2010), pp. 202–208

156. M. Worboys, M. Duckham, in *GIS: A Computing Perspective*, (CRC, 2004), ISBN: 0415283752

157. F. Bujor, E. Trouvé, L. Valet, J.-M. Nicolas, J.-P. Rudant, Application of log-cumulants to the detection of spatiotemporal discontinuities in multitemporal sar images. IEEE Trans. Geosci. Remote Sens. **42**(10), 2073–2084 (2004)

158. Y. Kosugi, M. Sakamoto, M. Fukunishi, W. Lu, T. Doihara, S. Kakumoto, Urban change detection related to earthquakes using an adaptive nonlinear mapping of high-resolution images. IEEE Geosci. Remote Sens. Lett. **1**(3), 152–156 (2004)

159. G. Di Martino, A. Iodice, D. Riccio, G. Ruello, A novel approach for disaster monitoring: fractal models and tools. IEEE Trans. Geosci. Remote Sens. **45**(6), 1559–1570 (2007)

160. R. Radke, S. Andra, O. Al-Kofahi, B. Roysam, Image change detection algorithms: a systematic survey. IEEE Trans. Image Process. **14**(3), 294–307 (2005)

161. R. Thoma, M. Bierling, Motion compensating interpolation considering covered and uncovered background. Sig. Process. Image Commun. **1**(2), 191–212 (1989)

162. T. Aach, A. Kaup, Bayesian algorithms for adaptive change detection in image sequences using markov random fields. Sig. Process. Image Commun. **7**(2), 147–160 (1995)

163. G. Chen, G.J. Hay, L.M. Carvalho, M.A. Wulder, Object-based change detection. Int. J. Remote Sens. **33**(14), 4434–4457 (2012)

164. B. Desclee, P. Bogaert, P. Defourny, Forest change detection by statistical object-based method. Remote Sens. Environ. **102**(1), 1–11 (2006)

165. J. Im, J. Jensen, J. Tullis, Object?based change detection using correlation image analysis and image segmentation. Int. J. Remote Sens. **29**(2), 399–423 (2008)

166. T. Aach, A. Kaup, R. Mester, Statistical model-based change detection in moving video. Sig. Process. **31**(2), 165–180 (1993)

167. E.J. Rignot, J.J. van Zyl, Change detection techniques for ERS-1 SAR data. IEEE Trans. Geosci. Remote Sens. **31**(4), 896–906 (1993)
168. J. Im, J. Jensen, A change detection model based on neighborhood correlation image analysis and decision tree classification. Remote Sens. Environ. **99**(3), 326–340 (2005)
169. Y. Yakimovsky, Boundary and object detection in real world images. J. ACM (JACM) **23**(4), 599–618 (1976)
170. D.H. Douglas, T.K. Peucker, Algorithms for the reduction of the number of points required to represent a digitized line or its caricature. Cartographica Int. J. Geograph. Inf. Geovisualization **10**(2), 112–122 (1973)
171. M. Kulldorff, W. Athas, E. Feurer, B. Miller, C. Key, Evaluating cluster alarms: a space-time scan statistic and brain cancer in los alamos, new mexico. Am. J. Public Health **88**(9), 1377–1380 (1998)
172. M. Kulldorff, Prospective time periodic geographical disease surveillance using a scan statistic. J. Roy. Stat. Soc. Ser. A (Stat. Soc.) **164**(1), 61–72 (2001)
173. D.J. Isaak, E.E. Peterson, J.M. Ver Hoef, S.J. Wenger, J.A. Falke, C.E. Torgersen, C. Sowder, E.A. Steel, M.-J. Fortin, C.E. Jordan et al., Applications of spatial statistical network models to stream data. Wiley Interdisc. Rev. Water **1**(3), 277–294 (2014)
174. D. Oliver, A. Bannur, J.M. Kang, S. Shekhar, R. Bousselaire, A k-main routes approach to spatial network activity summarization: A summary of results, in *2010 IEEE International Conference on Data Mining Workshops (ICDMW)* (IEEE, 2010), pp. 265–272
175. V.M.V. Gunturi, S. Shekhar, Lagrangian xgraphs: a logical data-model for spatio-temporal network data: A summary, in *Advances in Conceptual Modeling - ER 2014 Workshops, ENMO, MoBiD, MReBA, QMMQ, SeCoGIS, WISM, and ER Demos, Atlanta, GA, USA, October 27–29, 2014. Proceedings* (2014), pp. 201–211
176. V.M. Gunturi, E. Nunes, K. Yang, S. Shekhar, A critical-time-point approach to all-start-time lagrangian shortest paths: A summary of results, in *Advances in Spatial and Temporal Databases, vol. 6849*. Lecture Notes in Computer Science, ed. by D. Pfoser, Y. Tao, K. Mouratidis, M. Nascimento, M. Mokbel, S. Shekhar, Y. Huang (Springer, Berlin, 2011), pp. 74–91
177. V. Gunturi, S. Shekhar, K. Yang, A critical-time-point approach to all-departure-time lagrangian shortest paths. IEEE Trans. Knowl. Data Eng. **99**, 1 (2015)
178. S. Ramnath, Z. Jiang, H.-H. Wu, V.M. Gunturi, S. Shekhar, A spatio-temporally opportunistic approach to best-start-time lagrangian shortest path, in *International Symposium on Spatial and Temporal Databases* (Springer, 2015), pp. 274–291
179. J. Speed, Iot for v2v and the connected car, www.slideshare.net/JoeSpeed/aw-megatrends-2014-joe-speed
180. R.Y. Ali, V.M. Gunturi, A. Kotz, S. Shekhar, W. Northrop, Discovering non-compliant window co-occurrence patterns: A summary of results, in *Accepted in 14th International Symposium on Spatial and Temporal Databases* (2015)
181. ESRI, Breathe Life into Big Data: ArcGIS Tools and Hadoop Analyze Large Data Stores, http://www.esri.com/esriOnews/arcnews/summer13articles/breatheOlifeOintoObigOdata!
182. ESRI, ESRI: GIS and Mapping Software, http://www.esri.com
183. A. Aji, F. Wang, H. Vo, R. Lee, Q. Liu, X. Zhang, J. Saltz, Hadoop GIS: a high performance spatial data warehousing system over mapreduce. Proc. VLDB Endow. **6**(11), 1009–1020 (2013)
184. A. Eldawy, M.F. Mokbel, Spatialhadoop: a mapreduce framework for spatial data, in *Proceedings of the IEEE International Conference on Data Engineering (ICDE'15)* (IEEE, 2015)
185. C. Avery, *Giraph: Large-Scale Graph Processing Infrastructure on Hadoop* (Proceedings of the Hadoop Summit, Santa Clara, 2011)
186. Y. Low, J.E. Gonzalez, A. Kyrola, D. Bickson, C.E. Guestrin, J. Hellerstein, Graphlab: a new framework for parallel machine learning (2014). arXiv:1408.2041
187. G. Malewicz, M.H. Austern, A.J. Bik, J.C. Dehnert, I. Horn, N. Leiser, G. Czajkowski, Pregel: a system for large-scale graph processing, in *Proceedings of the 2010 ACM SIGMOD International Conference on Management of Data* (ACM, 2010), pp. 135–146

Part II
Classification of Earth Observation Imagery Big Data

Chapter 3
Overview of Earth Imagery Classification

Abstract This chapter overviews earth observation imagery big data and its general classification methods. We introduce different types of earth observation imagery big data and their societal applications. We also summarize some general classification algorithms. Open computational challenges are also identified in this area.

3.1 Earth Observation Imagery Big Data

With the advancement of remote sensing technology [1], a large repository of earth observation images is being collected at an increasing speed. A pixel in an image corresponds to a small area on the earth surface. Different spectral bands or layers in earth images measure the reflection of electromagnetic signal at different range of frequencies, such as visible spectrum (e.g., red, green, blue) or invisible spectrum (e.g., near infrared). The idea is that we can distinguish different land cover types on the earth surface based on their unique spectral signatures in reflection.

According to whether a remote sensor receives spectral reflection from signals emitted by itself or from signals emitted by the sun, remote sensing techniques are categorized into passive and active. In passive remote sensing, a remote sensor does not send out radiation signals itself, but collects reflection of the electromagnetic radiation of the earth surface from the sun. Thus, it may be impacted by the existence of cloud between the sensor and the land surface. In active remote sensing, a remote sensor sends out radiation signals and collects the reflection of the signal from the earth surface. Examples include radar and light detection and ranging (LiDAR) data. Active remote sensing is often less impacted by cloud since the signals can easily penetrate through cloud.

Earth observation imagery can also be categorized based on the platform on which the data is collected. Common platforms include satellite, space stations, aerial planes, and unmanned aerial vehicles (UAVs). Examples of satellite images include Moderate Resolution Imaging Spectroradiometer (MODIS) [2] and Landsat [3] from NASA as well as Sentinel-1 and Sentinel-2 from European Space Agency (ESA) [4]. MODIS imagery has a resolution of 200 m by 200 m, covering the entire globe around every other day. Landsat has higher spatial resolution (30 m) but

© Springer International Publishing AG 2017

Z. Jiang and S. Shekhar, *Spatial Big Data Science*,

DOI 10.1007/978-3-319-60195-3_3

covers the entire globe around every sixteen days. Satellite imagery has broad spatial coverage (often the entire globe), providing opportunities for large-scale (global or continental) analysis on the earth surface. There are also multiple spectral layers in satellite imagery. Aerial photos are taken by remote sensors on airplanes. Examples include high-resolution (meter to submeter) multi-spectral imagery (red, green, blue, near infrared) from National Agricultural Imagery Program (NAIP) [5] and civil air patrol imagery from NOAA National Geodetic Survey (NGS) [6]. The number of spectral bands is often small, but the spatial resolution is much higher than satellite images. The civil air patrol imagery are with very high spatial resolution (often inches) and provide true color image (red, green, and blue bands). The NGS data repository contains high-resolution true color aerial photos after major disasters such as floods and hurricanes in the USA. Compared with earth imagery from satellites, aerial imagery data has the advantages of avoiding clouds and providing very high spatial details. Moreover, individual agencies also collect LiDAR data for surface elevations to get terrain and topographic maps. Other aerial photos are collected from unmanned areal vehicles (UAVs). UAVs are often equipped with hyperspectral remote sensors collecting imagery with hundreds of spectral bands (very high spectral resolution) and high spatial resolution as well. Such hyperspectral imagery are increasingly important for precision agriculture, and forest health monitoring, particularly for data collection at a low altitude.

Many of existing earth observation imagery data such as Landsat, MODIS, Sentinel-1, and Sentinel-2 are now publicly available for free from the web site of USGS and ESA. Recently, Google developed a cloud-based platform called Google Earth Engine [7] that provides efficient support for accessing the tens of petabytes of earth imagery over the last four decades and analytics such as image processing and classification on a large scale. The system provides code development environment supporting JavaScript and Python and visualization on earth engine in the front end and runs the computational loads on clusters of machines on the back end. Google Earth image has been used to analyze the global forest cover change at relatively high resolution (30 m) using Landsat imagery [8]. This is the first time that forest cover change can be done globally at such high spatial resolution. Recently, people also use Google Earth Engine for surface water dynamic using Landsat imagery [9].

3.2 Societal Applications

Earth imagery big data provides unique opportunities to monitor the dynamics of the earth surface. There are broad applications that are closely related to the everyday life of billions of people. We list several examples of societal applications.

Water resource management: From poisonous drinking water in Toledo Ohio to catastrophic flood in Louisiana, water-related issues are costing the USA billions of dollars each year. Water challenge is considered among the greatest societal challenges in the twenty-first century. Earth observation imagery big data provides a unique opportunity to monitor surface water dynamics and water quality at large scale. For example, recently, research has been done to monitor the global water surface dynamics based on Landsat imagery (30 m by 3 m resolution) on Google Earth Engine. There has been also extensive works utilizing remote sensing imagery to study quality of inland and ocean water such as detecting algal blooms and chemical pollution [10]. For example, close to the shore of gulf of Mexico, there is a ocean dead zone without fish and animals due to algal blooms. Similarly, in the area of Great Lakes, algal blooms also cause toxic water for nearby cities. Earth image can provide early warning on water quality issues and monitor the progress of pollution [11].

Disaster management: Another important application is disaster management. In order to enhance situational awareness after a natural disaster, many agencies are collecting aerial photos after a disaster happens. These aerial photos usually can be collected right after disaster happens, faster than many satellite imagery and with much higher spatial resolution. For example, The National Geodetic Survey under NOAA has a repository of emergency response imagery after major hurricanes collected by organizations such as Civil Air Patrol [6]. The resolution of these imagery can be several inches, providing high spatial details for flood and damage. Similarly, high-resolution aerial photos can also be used to identify damaged areas after a tornado and automatically analyze the damage levels of buildings based on image analysis tools. This is important for decision makers in disaster management agencies.

Agriculture: Earth observation imagery can also be used in agricultural applications such as crop yield prediction and crop health monitoring. Monitoring crop health is important for food security. Particularly, with the development of hyperspectral remote sensors on unmanned aerial vehicles, new opportunities exist in providing agricultural plot management at fine spatial scales to improve crop yield while reducing cost of water and fertilizer. This is also called precision agriculture [12, 13]. Precision agriculture is very important because the energy, water, and food are often interconnected and interdependent [11, 14]. For example, food production involves consumption of water for irrigation and energy for fertilizer production, while energy generation in power plant consumes water resource for cooling systems. Earth observation data can play an important role in addressing the challenge of the nexus of food, energy, and water systems.

Land use and land cover modeling: Land use and land cover mapping [15] is a common application for earth observation imagery. Extensive work has been conducted to develop earth imagery classification methods in remote sensing community for this application. This is based on experiences that different land cover types on the earth surface show different spectral signatures in their reflected electromagnetic signals. Mapping the land cover of the earth surface is important for many reasons. For example, we can monitor the deforestation based on analyzing the change of forest coverage over time. We can also study environmental impacts of urbanization through analyzing patterns of land use change. Another specific example is wetland mapping (mapping the spatial distribution of different wetland types). Mapping wetland is an important application since wetlands are critical in purifying water, providing habitats for wild-life species, and buffering flood for disaster response.

3.3 Earth Imagery Classification Algorithms

Classifying earth observation imagery generally involves multiple steps. Raw earth imagery from satellite or airplanes often need to be first geo-referenced by adding in geographic coordinate such as latitude and longitude and rectified by reducing distortion due to the spherical earth surface. Many publicly available earth imagery datasets have already been geo-referenced and rectified. Second, the image may need to be preprocessed to remove cloud, noise, and shadows that can impact the classification performance. If supervised classification methods are used, ground truth class labels also need to be collected to create a training set. This can be done by sending out a field crew on the ground with GPS devices or hiring trained photo interpreters to manually label pixels. Then, a classification algorithm can be used to assign class labels to different image pixels. The results often need to be post-processed to enhance the quality of final data products. We focus more on the classification step with computer algorithms. Extensive research has been conducted on earth imagery classification algorithms. A survey on image classification can be found in [16]. There are different ways to categorize earth imagery classification techniques.

Supervised or unsupervised methods: Supervised methods learn a classification model based on training samples whose classes are already known. Training samples are often collected by sending a field crew on the ground with GPS devices or hiring visual interpreters to manually label image pixels. After a model is constructed, it can be applied to remaining pixels whose classes are unknown and to be predicted. Examples of supervised learning methods include maximum likelihood classifiers, decision tree, random forest, support vector machine, and neural network. For supervised methods, choosing appropriate and sufficient training samples is critical for performance. There are different ways to draw training samples such as random sample selection (selecting samples by random chance), stratified random sample (selecting samples from each class by random chance), clustered sampling (selecting a number of clusters of pixels), and systematic sampling (selecting samples from different cells of a regular grid so that samples cover most areas in the space) [17]. The bottom line is that training samples should be representative for the entire population. Unsupervised methods distinguish different classes of pixels based on their feature characteristics, such as clustering structure. Pixels are first clustered into several groups based on feature (spectral) similarity, and each group is assigned with a class label. One example is the K-means clustering method.

Per-pixel methods or object-based methods: Earth image classification methods can also be categorized based on the minimum classification unit. Many methods use pixels as the minimum classification units and thus are called per-pixel classifiers. Examples include the above-mentioned maximum likelihood classifiers [18], decision tree [19], random forest [20], support vector machine [21], and neural network [22]. These methods are traditionally developed for non-spatial data and thus ignore spatial autocorrelation and dependency across pixels. Their results tend to contain lots of artifacts (e.g., salt-and-pepper noise), particularly for high-resolution earth imagery with rich textures. This often requires tedious and time-consuming post-processing. In contrast to per-pixel methods, object-based image classification methods [23] first segment image pixels into homogeneous objects and classify these objects into different thematic classes. A popular tool for object-based image analysis is eCognition. Segmentation can be done both top down and bottom up. Parameters need to be specified to configure spatial scales of objects, and which similarity measure to use when merging pixels. The segmentation step often involves manual tuning by experts in order to get good results. After images are segmented into objects, we can extract features for objects such as mean and standard deviation of spectral values, shapes, and textures. These features can be used to learn a supervised classification model to automatically classify objects into different thematic classes. They can also be used to manually generate a set of decision rules to classify objects. This process, however, requires significant domain expertise on the spectral and spatial characteristics for different thematic classes. Object-based image analysis often shows significant accuracy improvements over per-pixel classifiers due to reduction of salt-and-pepper noise errors in prediction. However, it also involves manual tuning and is hard to scale to a large data.

Spectral classifier or spectral-spatial classifier: According to what features are used in the classification process, earth image classification methods can also be categorized into spectral classifiers and spatial–spectral classifiers. Spectral classifiers, as the name suggests, only use the spectral features of pixels. The idea is based on an assumption that different land surface types on the earth will show different signatures of spectral reflection. For example, vegetation cover often shows high reflection on near-infrared band, while water surface tends to have low reflection of near-infrared band. Since spectral properties are often considered on each pixel independently, these methods often produce salt-and-pepper noise. In contrast, spectral–spatial classifiers [24, 25] consider both the spectral properties of a pixel and the spatial contextual information. Examples of contextual variables include the textures of pixels in a window (e.g., homogeneity and heterogeneity). Spatial contextual information can also be incorporated in the post-processing steps, in which pixels can be smoothed based on the parcels where they are located in. Spectral–spatial classifiers can improve the pure spectral classifiers in reducing salt-and-pepper noise. In order to incorporate spatial contextual information, a large number of contextual features may need to be computed before running classification algorithms, making it computationally expensive [26].

Single classifier or ensemble of classifiers: Classifiers can be categorized by a single classifier or an ensemble of classifiers. Ensemble learning aims to learn a number of models to boost classification accuracy. The intuition is that if each classifier (expert) makes a slightly better prediction than random guess, voting by a large number of classifiers will have much better accuracy. Common ensemble methods include bagging [27], boosting [28], random forest [29], as well as decomposition-based ensemble such as mixture of experts [30]. For example, bagging method learns multiple base classifiers by manipulating learning samples (sampling with replacement) so that multiple training samples with different subsets can be drawn. In order to achieve good predictive performance, it has been suggested that different base classifiers should be as independent as possible. For instance, random forest enhances independence between base decision tree models by adding randomness to the feature selection process for decision tree nodes, i.e., only a small subset of features is randomly selected as candidates so that different trees have smaller chance to have same nodes. The advantage of learning an ensemble of classifiers is higher accuracy. However, ensemble method also impacts the interpretability of models and slows down the computational speed due to learning a large number of models.

Evaluation is an important step for earth image classification. The goal of evaluation is to provide a qualitative or quantitative summary of the classification performance of a method. There are different evaluation methods [17]. For example, we can evaluate a classification method based on the appearance of predicted class map. Method will have poor evaluation results if it produces lots of salt-and-pepper noise (also called poor appearance accuracy). Such maps cannot be used by decision-making agencies. Another more important evaluation strategy is quantitative evaluation. The idea is to have an independent set of learning samples (e.g., pixels) with class labels hidden as test sets. The true labels of test samples are compared against their predicted classes. Common quantitative metrics include confusion matrix, precision (user accuracy), recall (producer accuracy), and F-score. Another metric called Kappa is often used together, which measures how much better the classifier performs compared with random guess. Kappa is also used when comparing confusion matrices of two different classifiers to see if one's performance is statistically significantly better than another. Evaluating process for real-world earth image classification is often very challenging. First, there are often very small number of ground truth class labels due to limited budget. It is hard to get sufficient learning samples that can represent the entire population. Second, geographic data is heterogeneous in nature. Sample distribution in one location is often different from that in another location. Good evaluation results in one study area cannot guarantee similar performance in another different area. These issues are particularly important for large-scale (e.g., global or continental) earth imagery classification. In order to avoid overfitting, often simple and easily generalizable methods are used.

3.4   Generating Derived Features (Indices)

In earth image classification, we can use brightness of individual spectral bands (layers) as independent features to distinguish different thematic classes. For example, water pixels often correspond to extremely low values in the near-infrared band. In many applications, however, using individual spectral bands is insufficient to distinguish between different thematic classes. Derived features based on information from multiple spectral bands (or neighboring locations) are needed. In the past, people from different application domains have developed various derived features to solve a particular type of problems.

Vegetation indices are among the most successful examples of derived features in remote sensing [1]. The goal is to measure biomass and vegetation vigor based on brightness of spectral bands. High index value corresponds to large proportions of healthy vegetation within the pixel. In order to achieve this goal, band ratio between near-infrared (NIR) and red bands (R) are used. The effectiveness of this ratio is based on observations that in living vegetation, red light tends to be absorbed by chlorophyll but near-infrared light tends to be radiated by mesophyll tissue, so the ratio of near-infrared light over red light in the reflection is high for actively growing plants. In contrast, for non-vegetated surface such as water, building roof, bare soil, dead vegetation, the ratio of spectral responses on near infrared and red is not as significant. In order to make sure the value of the measure is between negative one and one, people have used *Normalized Digital Vegetation Index* (NDVI), $NDVI = \frac{NIR-R}{NIR+R}$. In Landsat imagery, NIR band is the fourth band, and R band is the third band. There are other vegetation indices such as Enhanced Vegetation Index (EVI), $EVI = G \times \frac{NIR-R}{NIR+C_1 \times R-C_2 \times B+L}$, where G and B are green and blue bands, C_1, C_2, and L are coefficients. EVI aims to improve sensitivity in high biomass regions and reduce the impact of soil background and atmospheric degradation. It is more sensitive to structural variation of vegetation canopy over time and is more effective in analyzing time series of vegetation level of a location to identify disturbance (e.g., forest fire). Vegetation indices are important in applications such as agriculture, e.g., monitoring crop health, predicting crop yield, understanding the phenology of plant species. The indices are also used to identify drought from historical earth imagery data. Other applications include land cover mapping and land cover change detection (e.g., deforestation).

Generating derived features has several advantages in earth image classification compared with using spectral band as independent features. First, the derived features can often be interpreted by the physical characteristics of surface spectral response. For example, the vegetation index NDVI can be interpreted by the photosynthetic effect in plants. Second, due to the fact that derived features are generated without relying on specific training labels, they are often more generalizable than supervised classification models. The reason is that supervised classification methods are often sensitive to the choice of training samples. The limitation of generating derived features or indices lies in the difficulties of identifying an appropriate index measure or function. There are multiple spectral bands that may contribute to the phenomena of interest, and these spectral bands may have a nonlinear compound effect. In machine learning, this is often called feature engineering. It is of interest to investigate how to utilize the supervised classification methods, e.g., decision trees, deep learning, to help identify critical derived features or indices for a specific phenomenon.

3.5 Remaining Computational Challenges

Spatial classification and prediction problems propose unique challenges as compared with traditional non-spatial classification and prediction problems due to the special characteristics of spatial data. First, spatial data is multi-dimensional in nature with various types of relationships (e.g., set-based, topological space, and metric space relationships) defined on it. This differs from non-spatial data such as numbers or characters. Moreover, from statistical and data analytics perspective, there are other challenges include spatial autocorrelation, spatial heterogeneity, limited ground truth, and multiple spatial scales and resolutions.

Spatial autocorrelation (dependency): In real-world spatial data, nearby locations tend to resemble each other, instead of being statistically independent. For example, the temperatures of two nearby cities are often very close. This phenomenon is also called the spatial autocorrelation effect. The existent of spatial autocorrelation is a challenge because it violates a common assumption by many classification and prediction models that learning samples are independent and identically distributed (i.i.d.). Ignoring the spatial autocorrelation effect may lead to poor prediction performance (e.g., errors or artifacts). Moreover, the extent of spatial dependency between locations may vary across different directions due to arbitrary geographic terrain and topographic features. In other words, spatial dependency is often none isotropic (anisotropic).

Spatial heterogeneity: Another unique characteristic of spatial data is that data distribution is often not identical in the entire study area, which is called spatial heterogeneity. For example, the same spectral signature in earth image pixels may correspond to different land cover types in different regions (also called spectral confusion). The same body language may have different meanings in different countries. Spatial heterogeneity is a challenge for classification and prediction problems because a global model learned from samples in the entire study area may perform poorly in a subregion.

Limited Ground Truth: Real-world spatial data often contains a large amount of information on explanatory features due to advancement in data collection techniques (e.g., remote sensor, location-based service), but limited ground truth information (e.g., target class labels). This is due to the fact that collecting ground truth for spatial data often involves sending a field crew to the ground or hiring well-trained experts to visually interpret the data, which is both expensive and time consuming. While limited ground truth is a shared challenge existing in other domain (e.g., web data) as well, the cost associated with ground truth collection often is somehow unique since it is related to travel costs on the ground.

Multiple Scales and Resolutions: The last major challenge in spatial classification and prediction is that data often exists in multiple spatial scales or resolutions. For example, in earth observation imagery, data resolutions range from submeter (high-resolution aerial photos), 30 m (Landsat satellite imagery), and 250 m (MODIS satellite). In precision agriculture, spatial data includes both ground sensor observations on soil properties at isolated points and aerial photos on the crop field for the entire area. This poses a challenge since many prediction methods often are developed for spatial data at the same scale or resolution.

Large data volume: As more and more earth observation imagery is being collected at an increasing speed, the data volume quickly exceeds the capability of traditional classification platforms. This is particularly important for global-scale analysis. Many current big data systems like Hadoop and Spark are not customized for spatial data. Big data platforms that are customized spatial data such as

SpatialHadoop only support simple spatial operations such as spatial joins across objects. Recently, Google Earth Engine, a cloud-based earth imagery processing platform, has been developed. It has been used for global-scale deforestation estimation and water surface mapping on Landsat imagery. But the algorithms used in these works are still non-spatial (e.g., decision trees).

Addressing these challenges requires the development of novel spatial classification algorithms. For example, in order to incorporate the spatial autocorrelation in traditional decision tree classification models, we need to change tree node tests or information gain heuristic. In the next several chapters, we will introduce several examples of recent works that address some of the above challenges, including spatial decision trees, and spatial ensemble learning. The goal is to illustrate how traditional non-spatial classification algorithms can be generalized for spatial data.

References

1. J.B. Campbell, R.H. Wynne, *Introduction to Remote Sensing*. (Guilford Press, 2011)
2. NASA. MODIS Moderate Resolution Imaging Spectroradiometer, https://modis.gsfc.nasa.gov/
3. United States Geological Survey, Landsat Missions, https://landsat.usgs.gov/
4. European Space Agency, The Copernicus Open Access Hub, https://scihub.copernicus.eu/
5. United States Department of Agriculture, National Agricultural Imagery Program, https://www.fsa.usda.gov/programs-and-services/aerial-photography/imagery-programs/naip-imagery/
6. National Oceanic and Atmospheric Administration, National Geodetic Survey, https://www.ngs.noaa.gov/
7. Google Earth Engine Team, Google earth engine: A planetary-scale geo-spatial analysis platform, https://earthengine.google.com (2015)
8. M.C. Hansen, P.V. Potapov, R. Moore, M. Hancher, S. Turubanova, A. Tyukavina, D. Thau, S. Stehman, S. Goetz, T. Loveland et al., High-resolution global maps of 21st-century forest cover change. Science **342**(6160), 850–853 (2013)
9. J.-F. Pekel, A. Cottam, N. Gorelick, A.S. Belward, High-resolution mapping of global surface water and its long-term changes. Nature **540**, 418–422 (2016)
10. T. Kutser, L. Metsamaa, N. Strömbeck, E. Vahtmäe, Monitoring cyanobacterial blooms by satellite remote sensing. Estuar. Coast. Shelf Sci. **67**(1), 303–312 (2006)
11. E. Eftelioglu, Z. Jiang, X. Tang, S. Shekhar, The nexus of food, energy, and water resources: Visions and challenges in spatial computing. in *Advances in Geocomputation*. (Springer, Berlin, 2017), pp. 5–20
12. G. Ruß, A. Brenning, Data mining in precision agriculture: management of spatial information. in *Computational Intelligence for Knowledge-Based Systems Design*. (Springer, Berlin, 2010), pp. 350–359
13. C. Zhang, J.M. Kovacs, The application of small unmanned aerial systems for precision agriculture: a review. Precis. Agric. **13**(6), 693–712 (2012)
14. E. Eftelioglu, Z. Jiang, R. Ali, S. Shekhar, Spatial computing perspective on food energy and water nexus. J. Environ. Stud. Sci. **6**(1), 62–76 (2016)
15. A. Karpatne, Z. Jiang, R.R. Vatsavai, S. Shekhar, V. Kumar, Monitoring land-cover changes: A machine-learning perspective. IEEE Geosci. Rem. Sens. Mag. **4**(2), 8–21 (2016)
16. D. Lu, Q. Weng, A survey of image classification methods and techniques for improving classification performance. Int. J. Rem. Sens. **28**(5), 823–870 (2007)
17. R.G. Congalton, A review of assessing the accuracy of classifications of remotely sensed data. Rem. Sen. Environ. **37**(1), 35–46 (1991)

18. A. Strahler, The use of prior probabilities in maximum likelihood classificaiton of remote sensing data. Rem. Sens. Environ. **10**, 135–163 (1980)
19. M.A. Friedl, C.E. Brodley, Decision tree classification of land cover from remotely sensed data. Rem. Sens. Environ. **61**(3), 399–409 (1997)
20. M. Pal, Random forest classifier for remote sensing classification. Int. J. Rem. Sens. **26**(1), 217–222 (2005)
21. F. Melgani, L. Bruzzone, Classification of hyperspectral remote sensing images with support vector machines. IEEE Trans. Geosci. Rem. Sens. **42**(8), 1778–1790 (2004)
22. J.A. Benediktsson, P.H. Swain, O.K. Ersoy, Neural network approaches versus statistical methods in classification of multisource remote sensing data. IEEE Trans. Geosci. Rem. Sens. **28**, 540–552 (1990)
23. G. Hay, G. Castilla, Geographic object-based image analysis (geobia): A new name for a new discipline. in *Object-Based Image Analysis*. (Springer, Berlin, 2008), pp. 75–89
24. Y. Tarabalka, J.A. Benediktsson, J. Chanussot, Spectral-spatial classification of hyperspectral imagery based on partitional clustering techniques. IEEE Trans. Geosci. Rem. Sens. **47**(8), 2973–2987 (2009)
25. M. Fauvel, Y. Tarabalka, J.A. Benediktsson, J. Chanussot, J.C. Tilton, Advances in spectral-spatial classification of hyperspectral images. Proc. IEEE **101**(3), 652–675 (2013)
26. J.A. Benediktsson, J.A. Palmason, J.R. Sveinsson, Classification of hyperspectral data from urban areas based on extended morphological profiles. IEEE Trans. Geosci. Rem. Sens. **43**(3), 480–491 (2005)
27. L. Breiman, Bagging predictors. Mach. Learn. **24**(2), 123–140 (1996)
28. Y. Freund, R. Schapire, N. Abe, A short introduction to boosting. J. Jpn. Soc. Artif. Intell. **14**(771–780), 1612 (1999)
29. L. Breiman, Random forests. Mach. Learn. **45**(1), 5–32 (2001)
30. S.E. Yuksel, J.N. Wilson, P.D. Gader, Twenty years of mixture of experts. IEEE Trans. Neural Netw. Learn. Syst. **23**(8), 1177–1193 (2012)

Chapter 4
Spatial Information Gain-Based Spatial Decision Tree

Abstract This chapter introduces one specific technique of earth imagery classification called spatial information gain-based spatial decision tree. The corresponding spatial decision tree learning algorithm incorporates spatial autocorrelation effect by a new spatial information gain (SIG) measure.

4.1 Introduction

Given learning samples from a spatial raster dataset, the geographical classification problem aims to learn a decision tree classifier that best characterizes the relationship between the explanatory features and ground truth class labels in the data, namely minimizing classification errors and salt-and-pepper noise. Once a decision tree classifier is learned, it can be applied to future datasets where the ground truth class labels are unknown and need to be predicted. Figure 4.2 illustrates a decision tree classifier that was learned from the raster dataset in Fig. 4.1. The figure shows three explanatory features of the data (Fig. 4.1a–c) and the target ground truth class label (Fig. 4.1d) where pixels in the upper half are selected as learning samples. The decision tree classifier learned based on these samples further produces a classification of the entire area, as shown in Fig. 4.1e.

4.1.1 Societal Application

The geographical classification problem is important for many applications ranging from land cover classification [1] to lesion classification in medical diagnosis [2, 3], etc.

For example, in the remote sensing community, geographical classification techniques are used for wetland mapping, namely classifying wetland and upland cover types based on explanatory features, including spectral information and topographical derivatives (e.g., slope and curvature). This is critically important for a broad range of natural resource management concerns, including regulatory purposes, monitoring of changes in extent or function due to natural and man-made causes;

© Springer International Publishing AG 2017
Z. Jiang and S. Shekhar, *Spatial Big Data Science*,
DOI 10.1007/978-3-319-60195-3_4

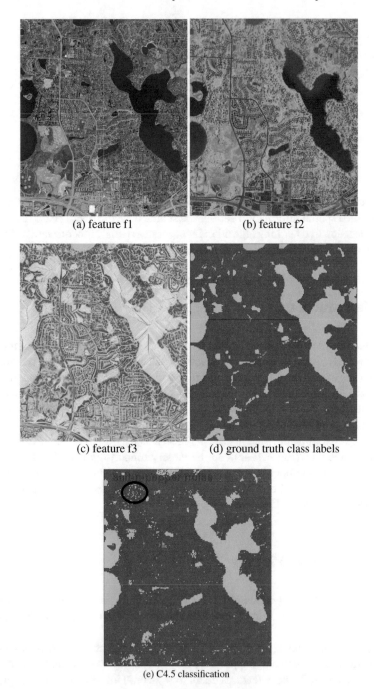

(a) feature f1   (b) feature f2

(c) feature f3   (d) ground truth class labels

(e) C4.5 classification

Fig. 4.1 Example of a raster dataset, learning samples from upper half, in **d** and **e**: *red* for dryland, *green* for wetland (best viewed in color)

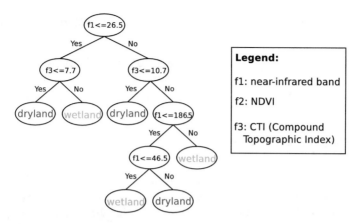

Fig. 4.2 Output decision tree by C4.5 (best viewed in color)

and area selection with the most suitable hydrologic and vegetative characteristics for wetland restoration or conservation [4–6]. In climate science domain, studies show that wetlands contribute to over half of the global emissions of methane, a powerful greenhouse gas; and accurate wetland mapping thus helps climate scientists to understand global warming process [7].

4.1.2 Challenges

The geographical classification problem is challenging due to the existence of spatial autocorrelation structure arising from phenomena such as "patches." For example, in wetland mapping applications, ground truth class labels (e.g., wetland) often form geographical "patches" (contiguous areas of the same class). In high-resolution remote sensing images, the pixel size is much smaller than the patch size. Thus, a geographically contiguous patch may have many pixels with the same ground truth class label, leading to high spatial autocorrelation. For instance, in Fig. 4.1d, a wetland patch in green color is circled, which consists of hundreds of pixels. The whole ground truth class label map in Fig. 4.1d has a value of 0.97 for Moran's I (with queen neighborhood), indicating high spatial autocorrelation.

Failing to incorporate spatial autocorrelation, i.e., assuming identical independent distribution, will impact classification accuracy and often result in salt-and-pepper noise. For example, Fig. 4.1e is the prediction of traditional C4.5 decision tree learning algorithm, containing a lot of salt-and-pepper noise (circled in top left corner), compared with ground truth class labels in Fig. 4.1d.

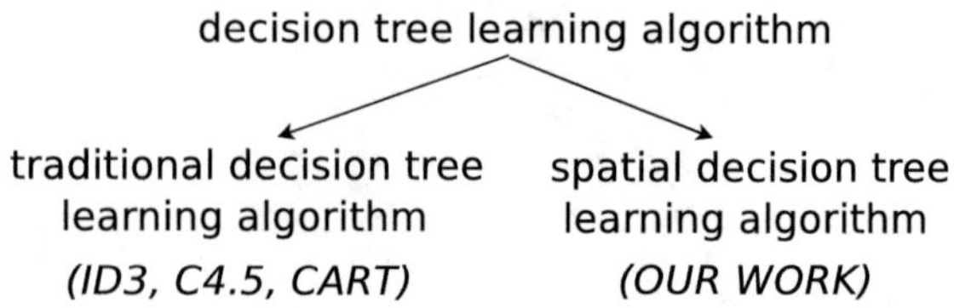

Fig. 4.3 Related work
categorization

decision tree learning algorithm

traditional decision tree   spatial decision tree
learning algorithm   learning algorithm
(ID3, C4.5, CART)   *(OUR WORK)*

4.1.3   Related Work Summary

Traditional decision tree learning algorithms include ID3 [8], C4.5 [9] and CART
[10], as shown in Fig. 4.3. When applied to geographical classification, traditional
decision tree learning algorithms implicitly assume the data items are independent,
ignoring spatial autocorrelation effect. Thus, the classification result contains salt-
and-pepper noise (as shown in Fig. 4.1e). In contrast, this chapter introduces our
recent spatial autocorrelation aware spatial decision tree learning algorithm [11] as
shown in Fig. 4.3. Compared with traditional decision tree learning algorithm, our
spatial decision tree aims to reduce salt-and-pepper noise, which may also improve
classification
accuracy.

Contributions: This chapter makes the following contributions: (a) we propose a
new spatial decision tree (SDT) model, including a spatial information gain interestingness measure; (b) we develop an SDT learning algorithm; and (c) we conduct a
case study to evaluate the proposed approach on a real-world remote sensing dataset.

Scope: This chapter focuses on decision tree classifiers. Other geographical classification techniques such as Bayesian classifier, SVM, neural network, logistic regression, GEOBIA (geographical object-based image analysis) fall outside the scope of
the study. This chapter does not consider ensembles of decision trees, e.g., bagging,
boosting. The choice of sampling method on spatial dataset is also beyond the scope
of the present research. Finally, this chapter does not address heterogeneity, a second
important characteristic of spatial data. This topic will be explored in the future work.

Outline: The rest of this chapter is organized as follows: Sect. 4.2 formalizes
the problem. Section 4.3 defines some basic concepts and introduces the proposed
spatial decision tree learning algorithm. Section 4.4 describes a case study on a real-
world remote sensing dataset. Computational performance tuning and experimental
evaluation are presented in Sect. 4.5. Section 6 concludes this chapter with some
discussion and possible future work.

4.2   Problem Formulation

In this section, we formulate the spatial decision tree learning problem in geographical classification. We first provide a formal problem definition and then illustrate the
problem inputs and outputs with a first principle example.

Formally, the problem may be defined as follows:

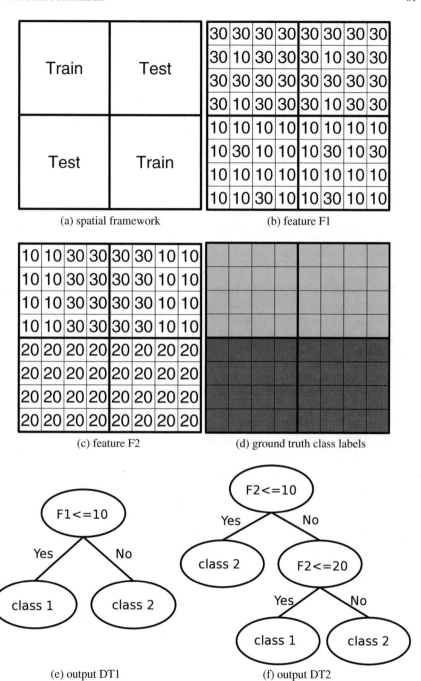

(a) spatial framework

(b) feature F1

(c) feature F2

(d) ground truth class labels

(e) output DT1

(f) output DT2

Fig. 4.4 Example of problem input and output with *red* for class 1 and *green* for class 2 (best viewed in color)

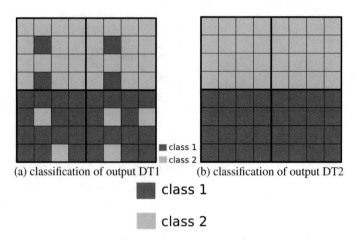

(a) classification of output DT1   (b) classification of output DT2

■ class 1

■ class 2

Fig. 4.5 Classification results of decision tree classifiers from Fig. 4.4e, f

Given: A spatial framework s, learning samples (i.e., training, test samples) located in s, each sample specifies explanatory features and a class label
Find: a decision tree classifier based on training set
Objective: minimize classification error on test set
Constraint:

(a): spatial framework is a regular grid
(b): spatial autocorrelation exists in ground truth class labels

   Example: Figure 4.4 shows an example of problem input and output. Figure 4.4a has four subareas: the upper left and lower right subareas contain training pixels, and the others contain test pixels. Consider two explanatory features, namely $F1$ and $F2$, whose values are the numbers projected on pixels in Fig. 4.4b, c, respectively. Ground truth class labels are shown in Fig. 4.4d. Two candidate output decision tree (DT) classifiers, namely DT1 and DT2, are shown in Fig. 4.4e, f. The corresponding classification results of DT1 and DT2 are in Fig. 4.5a, b, respectively. We can see that DT2 has less salt-and-pepper noise and better classification accuracy than DT1, referring to the ground truth class labels in Fig. 4.4d.
   Discussion: The ground truth class labels in Fig. 4.4d form contiguous "patches" spanning many pixels and show strong spatial autocorrelation (as discussed in Sect. 1.1 societal application). A pixel classified differently from all its neighbors within a patch is considered as salt-and-pepper noise and is less desirable. For example, in Fig. 4.5a, green pixels within the red patch in the lower half are salt-and-pepper noise.

4.3 Proposed Approach

This section introduces the proposed spatial decision tree learning algorithm, which is designed to maintain the spatial autocorrelation structure of ground truth class labels and resist salt-and-pepper noise. First, we explain some basic concepts and define our interestingness measure. We then describe the spatial decision tree learning algorithm with its pseudocode and an execution trace.

4.3.1 Basic Concepts

Spatial Neighborhood Relationship: A neighborhood relationship is used to characterize the range of spatial dependency between data items at nearby locations. There are different ways to determine neighbors, such as Euclidean distance based (a threshold or k-nearest neighbor in point reference data) and topographical relationship based (touch in areal data model). A spatial neighborhood relationship can be

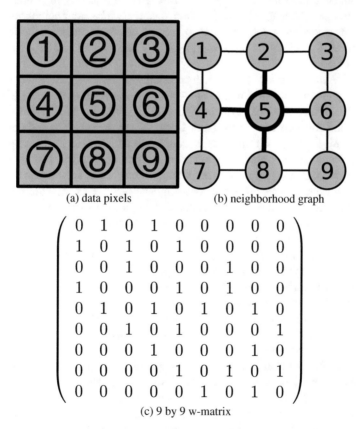

(a) data pixels (b) neighborhood graph

$$\begin{pmatrix} 0 & 1 & 0 & 1 & 0 & 0 & 0 & 0 & 0 \\ 1 & 0 & 1 & 0 & 1 & 0 & 0 & 0 & 0 \\ 0 & 0 & 1 & 0 & 0 & 0 & 1 & 0 & 0 \\ 1 & 0 & 0 & 0 & 1 & 0 & 1 & 0 & 0 \\ 0 & 1 & 0 & 1 & 0 & 1 & 0 & 1 & 0 \\ 0 & 0 & 1 & 0 & 1 & 0 & 0 & 0 & 1 \\ 0 & 0 & 0 & 1 & 0 & 0 & 0 & 1 & 0 \\ 0 & 0 & 0 & 0 & 1 & 0 & 1 & 0 & 1 \\ 0 & 0 & 0 & 0 & 0 & 1 & 0 & 1 & 0 \end{pmatrix}$$

(c) 9 by 9 w-matrix

Fig. 4.6 Examples of rook neighborhood graph, w-matrix. Number 1–9 is data item id, color is class label

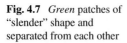

Fig. 4.7 *Green* patches of "slender" shape and separated from each other

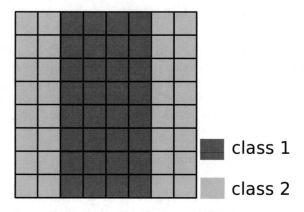

represented as edges in a *neighborhood graph*, whose vertices are spatial data items (e.g., points, raster grid, areal polygons). The relationship can also be presented as an adjacency matrix, often called a *w-matrix*. Figure 4.6a shows an example of raster framework containing nine pixels with ids 1–9. Under a rook neighborhood relationship, the neighborhood graph is in Fig. 4.6b, e.g., pixel 5 has four neighbors (2, 4, 6, 8). The corresponding w-matrix is shown in Fig. 4.6c. For instance, in the 5th row of the w-matrix, four elements ($w_{5,2}$, $w_{5,4}$, $w_{5,6}$, $w_{5,8}$) are 1s, indicating four neighbors (2,4,6,8) of pixel 5.

Spatial Autocorrelation: Spatial autocorrelation statistics measure the degree of dependency among non-spatial attribute values of neighboring data items. There are various definitions of spatial autocorrelation [12], including both global and local versions, and for both categorical (e.g., Mental's M2, joint count) and continuous (e.g., Moran's I, Geary's C) variables. In addition, Xiang Li et al. [13] have proposed a distance-based autocorrelation measure for decision tree models, and more specifically, autocorrelation is high when the homogeneous points are close to each other and heterogeneous points are far away from each other. This measure reflects the global clusterness of homogeneous points well, but it favors single spherical homogeneous cluster or patch. In applications such as land cover classification where homogeneous "patches" are of arbitrary shape (e.g., "slender" shapes) or separated far away from each other (e.g., separate lakes as wetlands), the average distance between homogeneous points will be so large that the actual autocorrelation effect is underestimated. For example, Fig. 4.7 shows two classes with high spatial autocorrelation. However, since the two green patches are slender in shape and separated far away from each other, the average distance between green pixels is quite high (6.54 units in Euclidean distance), and a distance-based measure falsely indicates low spatial autocorrelation.

To reflect the fact that in the real-world spatial autocorrelation is significant at the local neighborhood level, we introduce an autocorrelation measure named the *Gamma index* [12]. The detailed formula is given in equation(1) where i, j is any pair of spatial data items; $a_{i,j}$ and $b_{i,j}$ are spatial similarity and class similarity, respectively. Here, spatial similarity is defined as elements of w-matrix, $w_{i,j}$, and class similarity is defined as an indicator function $\delta_{i,j}$ (value is 1 when i, j are in the same class; and 0 otherwise). Thus, the Gamma index is the total count of homogeneous neighbors for all items ($c(i)$ is the count for item i). For example, given rook neighborhood relationship, the Gamma index of Fig. 4.5a is 149, while the index of Fig. 4.5b is 184.

$$\Gamma = \sum_{i,j} a_{i,j} b_{i,j} = \sum_{i,j} w_{i,j} \delta_{i,j} = \sum_{i} c(i) \tag{4.1}$$

A local version of this autocorrelation measure named *local Gamma* is shown in equation (2). For example, in previous Fig. 4.6b, pixel 5 has four homogeneous neighbors, making its local gamma autocorrelation value 4.

$$\Gamma_i = \sum_{j} a_{i,j} b_{i,j} = \sum_{j} w_{i,j} \delta_{i,j} = c(i) \tag{4.2}$$

Entropy and Information Gain (IG): In the training or decision tree induction phase, learning samples are split by a candidate feature and test threshold (e.g., $F \leq 10$) within a tree node. The goal is to reduce the impurity of class distributions as much as possible, i.e., ideally, all learning samples from one class are below the threshold, and the other class is above the threshold. *Entropy* is a measure of the impurity of class probability distribution and *information gain* is the decrease of entropy after a split. Formally,

$$Entropy = -\sum_{l} p_l \log p_l \tag{4.3}$$

where p_l is the probability of class l. The entropy after split, $E\prime$, is the average of entropy in each split result, shown as follows:

$$E\prime = \frac{n_1}{n_1 + n_2} E_1 + \frac{n_2}{n_1 + n_2} E_2 \tag{4.4}$$

where n_1, n_2 are the number of learning samples in each split result. Information gain is the decrease of entropy, defined as:

$$IG = E - E\prime \tag{4.5}$$

For example, in Fig. 4.9a, the top box shows the learning samples (as circles) in two classes (i.e., class 1 in red, class 2 in green). The entropy before the split is $-1/2 \log 1/2 - 1/2 \log 1/2 = 1$. After the split, learning samples below the threshold are shown as diamonds; and those above the threshold are shown as squares. The entropy after the split is $(-2/15 * \log 2/15 - 13/15 \log 13/15) * 15/32 + (-14/17 \log 14/17 - 3/17 \log 3/17) * 17/32 = 0.623$. So the traditional information gain is $1 - 0.623 = 0.377$.

Neighborhood Split Autocorrelation Ratio (NSAR): A split on spatial learning samples not only changes the impurity of the probability distribution (i.e., entropy) of ground truth class labels but also influences their spatial distribution (i.e., spatial

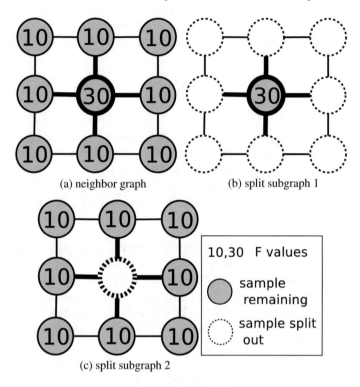

(a) neighbor graph   (b) split subgraph 1

(c) split subgraph 2

Fig. 4.8 Examples of neighborhood split $F \leq 10$ and NSAR value

autocorrelation level). NSAR measures the decrease of the local gamma autocorrelation level after a neighborhood of learning samples is split. Formally,

$$NSAR_i = \frac{\Gamma_i^{after}}{\Gamma_i^{before}} = \frac{\sum_j w\prime_{i,j}\delta_{i,j}}{\sum_j w_{i,j}\delta_{i,j}} = \frac{c\prime(i)}{c(i)} \tag{4.6}$$

where $w\prime_{i,j}$ is w-matrix elements after a split, and $c\prime(i)$ is the count of homogeneous neighbors after a split. Thus, NSAR is the count of homogeneous neighbors of a sample after a split over the count before the split. Its value ranges between 0 and 1 (value is 1 when $c(i) = 0$). The higher the NSAR, the less salt-and-pepper noise a split makes. Figure 4.8 shows an example of a split of a neighborhood graph. Before the split in (a), the central sample (with F value 30) has four homogeneous neighbors. However, after the split by $F <= 10$, no homogeneous neighbors remain in the same subgraph (shown in subfigure (b)). Thus, the NSAR value of the central sample is $0/4 = 0$. Another example is the green diamonds in central box of Fig. 4.9a. As can be seen, if a split breaks down the spatial autocorrelation structure and creates salt-and-pepper noise, the NSAR value will be low.

Proposed Interestingness Measure: Spatial Information Gain (SIG). SIG is a balance between reduction of class impurity and maintenance of spatial autocorrelation structure (i.e., resistance to salt-and-pepper noise). More specifically, it is a weighted

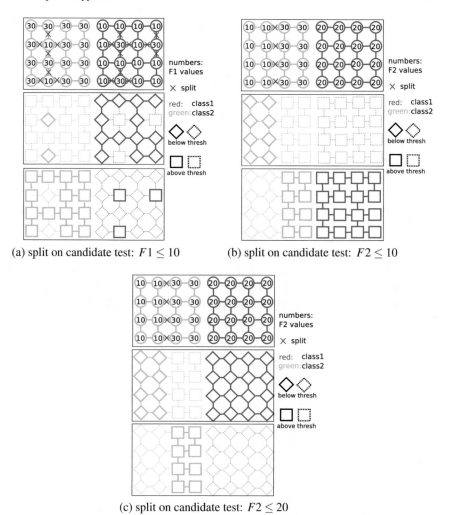

(a) split on candidate test: $F1 \leq 10$ (b) split on candidate test: $F2 \leq 10$

(c) split on candidate test: $F2 \leq 20$

Fig. 4.9 Comparison of three candidate tree features and thresholds for root node

sum of traditional information gain (IG) and average NSAR over all local neighborhoods (denoted as $\overline{NSAR}$). Formally,

$$SIG = (1 - \alpha)IG + \alpha\overline{NSAR} = (1 - \alpha)IG + \alpha\frac{1}{m}\sum_{k=1}^{m}NSAR(k) \qquad (4.7)$$

where k is a neighborhood index, varying from 1 to m.

The balancing parameter α could be learned from data, using an independent validation set (as discussed in the case study in Sect. 4.4). For instance, if the learning samples follow an identical independent distribution, then the α value will be close to zero, and SIG turns into traditional IG.

4.3.2 Spatial Decision Tree Learning Algorithm

The spatial decision tree learning algorithm consists of two phases: Phase I generates a neighborhood graph from training pixels and then constructs a spatial decision tree model; Phase II makes predictions (i.e., classification) on test samples based on the constructed spatial decision tree. The training phase is shown in Algorithm 1. The prediction phase is the same as traditional decision tree algorithms.

SDT-Train: The SDT-Train algorithm (Algorithm 1) takes a neighborhood graph of training pixels with feature values and ground truth class labels, SIG balancing parameter α, and minimum tree node size as inputs. It then constructs a spatial decision tree model. Similar to traditional decision tree learning algorithms such as C4.5, SDT-Train is also a top-down divide-n-conquer algorithm with a greedy strategy. But it differs from traditional decision tree learning algorithms in two aspects: firstly, it maintains a neighborhood graph data structure of current training samples, which will be split into two subgraphs based on a selected feature and threshold; secondly, in order to evaluate a candidate test feature and split threshold, it uses the spatial information gain interestingness measure, which reflects not only the reduction of class impurity but also the maintenance of autocorrelation structure (i.e., resistance to salt-and-pepper noise). Now, we introduce the SDT-Train algorithm step by step:

Algorithm 1 SDT-Train($\mathcal{G}$,$\mathcal{F}$,$\mathcal{C}$, α,c_0)

Input:
- $\mathcal{G}$: neighborhood graph of training pixels
- $\mathcal{F}$: feature set of neighborhood graph nodes
- $\mathcal{C}$: class label set of neighborhood graph nodes
- α: weight for autocorrelation term in SIG measure
- c_0: minimum decision tree node size

Output:
- root of a spatial decision tree model

1: **if** $|\mathcal{G}| < c_0$ or $\mathcal{G}.\mathcal{C}$ in the same class **then**
2:    Create a leaf node $\mathcal{L}$ from $\mathcal{G}$
3:    **Return** $\mathcal{L}$
4: **for each** feature $f \in \mathcal{F}$ **do**
5:    Sort f values of all $\mathcal{G}$ nodes in ascending order, as $\mathcal{G}.f$
6:    **for each** distinct value δ of $\mathcal{G}.f$ **do**
7:        Create candidate test $f \leq \delta$
8:        Scan and split $\mathcal{G}$ into $\mathcal{G}_1$ and $\mathcal{G}_2$ based on $f \leq \delta$
9:        Calculate IG, $\overline{NSAR}$
10:       $SIG = (1 - \alpha)IG + \alpha\overline{NSAR}$
11: Find candidate test $f_0 \leq \delta_0$ with largest SIG
12: Create internal node $\mathcal{I}$ with test $f_0 \leq \delta_0$
13: Split $\mathcal{G}$ into $\mathcal{G}_1$ and $\mathcal{G}_2$ based on f_0 and δ_0
14: $\mathcal{I}$.LeftChild=**SDT-Train**($\mathcal{G}_1$, $\mathcal{F}$, $\mathcal{C}$, α, c_0)
15: $\mathcal{I}$.RightChild=**SDT-Train**($\mathcal{G}_2$, $\mathcal{F}$, $\mathcal{C}$,α, c_0)
16: **Return** $\mathcal{I}$

Steps 1–4 check the stop criteria: if all the nodes or training pixels in the neighborhood graph are in the same class or the node number is smaller than the minimum decision tree node size, then a leaf labeled with the majority class is returned.

Steps 5–13 evaluate the SIG measure of all candidate features and possible splitting thresholds. More specifically, step 6 does a quick sort on all values of a feature f, and each distinct feature value becomes a candidate test threshold δ. Steps 9 and 10 scan and split the neighborhood graph based on candidate test $f \leq \delta$ and compute the SIG.

Steps 14–16 make a greedy choice. The feature and threshold with the largest SIG are selected, an internal node is created, and the neighborhood graph is then split into two subgraphs. This is the "divide" part.

Steps 17–18 call the SDT-Train function recursively on each subgraph. This is the "conquer" part.

4.3.3 An Example Execution Trace

In this subsection, we illustrate the execution trace of the SDT-Train and SDT-Predict algorithms. The example input remains as shown in Fig. 4.4a–d. We have 32 training samples (in upper left and lower right corners) with features F1 and F2, and ground truth class labels "class1" in red and "class2" in green. In the learning algorithm, SIG balancing parameter α is set to 0.5; the minimum tree node size is 4; the spatial neighborhood relationship is a rook neighborhood, i.e., a pixel has four neighbors sharing a boundary with it or "touching" it.

Training Phase

Neighborhood graph generation: Based on the 32 training pixels from Fig. 4.4 and the rook neighborhood relationship, the generated neighborhood graph is shown in the top most subfigure of Fig. 4.9a–c. Colors are ground truth class labels, and numbers are feature values of F1 and F2, respectively.

Steps 1–4 check the stop criteria. Since the neighborhood graph size 32 is above the minimum threshold 4, and nodes of the training pixels are not in the same class, the stop criteria is not satisfied.

Steps 5–13 evaluate the spatial information gain (SIG) of all candidate features and split thresholds. For example, for feature F1, all the values are sorted at first. There are two distinct values, i.e., 10 and 30. Thus, the candidate test threshold is $F1 \leq 10$. (Note that $F1 \leq 30$ is excluded since it will not split the graph into two subgraphs and thus not purify the classes.) Similarly, F2 has distinct values 10, 20, and 30, so the candidate tests on F2 are $F \leq 10$ and $F2 \leq 20$.

Fig. 4.9a–c shows the three candidate tests and the neighborhood graph splits based on each test. The traditional information gain and spatial information gain are given in Table 4.1.

Step 14–16 make a greedy choice. According to Table 4.1, a traditional decision tree will select $F1 \leq 10$ since it purify class distribution to the most, i.e., maximizing IG. In contrast, our spatial decision tree will select $F2 \leq 10$ or $F2 \leq 20$, which

Table 4.1 Comparison of IG and SIG of three candidate tests

	$F_1 \leq 10$	$F_2 \leq 10$	$F_2 \leq 20$
IG	**0.377**	0.311	0.311
$\overline{NSAR}$	0.669	0.927	0.927
SIG	0.503	**0.619**	**0.619**

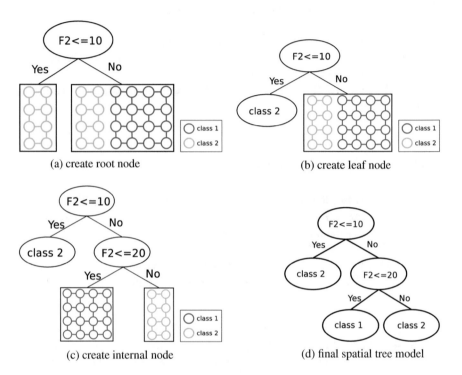

Fig. 4.10 Example of execution trace

not only purifies class distribution but also maintains the spatial autocorrelation level among the training pixel nodes. Without loss of generalizability, we select $F2 \leq 10$, create an internal tree node, and split the graph into two subgraphs, as shown in Fig. 4.10a.

Steps 17–18 call SDT-Train on each subgraph recursively. For example, the left subgraph is in same class, so a leaf node is returned as shown in Fig. 4.10b. For the right subgraph, similar to previous steps, the best test $F2 \leq 20$ is selected as in (c). Subfigure (d) shows the final spatial decision tree constructed, the same as Fig. 4.4f.

4.4 Evaluation

We evaluated the proposed spatial decision tree learning algorithm in a case study on a real-world remote sensing dataset. The goal of the case study was to answer the following questions:

- Does incorporating spatial autocorrelation effect into decision tree learning algorithm help to improve classification accuracy?
- Does incorporating spatial autocorrelation effect into decision tree learning algorithm help to reduce salt-and-pepper noise in classification?
- How may one choose α, the balancing parameter for the spatial information gain interestingness measure?

4.4.1 Dataset and Settings

Dataset description: The real-world high-resolution (3m by 3m) remote sensing dataset was collected from the city of Chanhassen, Minnesota (shown in Fig. 4.11). Explanatory features consisted of multi-temporal (2003, 2005, 2008) spectral information (e.g., R, G, B, NIR bands of aerial photographs) and topographical derivatives (e.g., slope and curvature). Ground truth class labels (wetland and upland cover types) were depicted by wetland delineation field crew and trained photointerpreters. The training area was in the central east, and the test area was in the northeast.

Learning Samples Creation: Training samples and validation samples were generated by systematic stratified cluster sampling. Firstly, a coarse grid was drawn over the training area and each coarse cell (98 pixel by 98 pixel) was a primary sample unit (PSU). Then, we randomly selected eight dryland PSUs and eight wetland PSUs for training, two dryland PSUs and two wetland PSUs for validation, identified by the majority class. Within each PSU, we further randomly selected 1000 pixels. Thus, training set consists of 16000 pixels and validation set consists of 4000 pixels. *Test samples* consists of the pixels in the entire test area.

Parameter settings: The distance threshold for neighborhood relationship between data samples was 18 m (6 pixels). Neighbor-wise tree node test ratio θ was 0.6. Minimum tree node size c_0 was 100 pixels. The balancing parameter α in SIG was 0.26 (details are in Sect. 4.4).

Evaluation candidates:

(1) the traditional C4.5 decision tree learning algorithm (DT)
(2) the spatial decision tree learning algorithm (SDT)

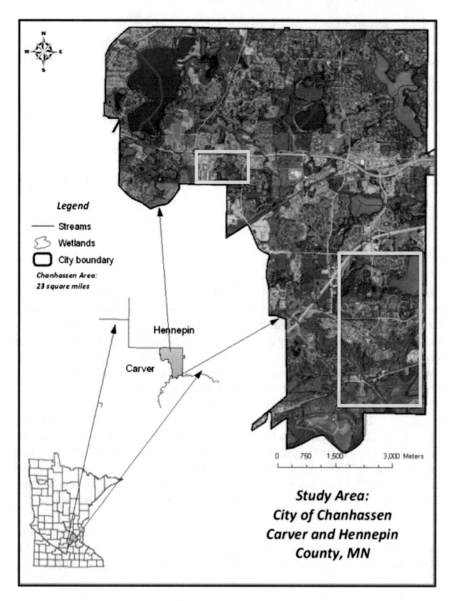

Fig. 4.11 Study area: Chanhassen, MN (best viewed in color) Courtesy of Lian Rampi

Table 4.2 Confusion matrix of DT

	Classified dry	Classified wet
Truth dry	826,844	**47,173**
Truth wet	**26,385**	188,814

Table 4.3 Confusion matrix of SDT

	Classified dry	Classified wet
Truth dry	836,961	**37,056**
Truth wet	**23,565**	191,634

4.4.2 Does Incorporating Spatial Autocorrelation Improve Classification Accuracy?

Classification accuracy of the two candidate algorithms on test area is shown in the confusion matrices in Tables 4.2 and 4.3. In a confusion matrix, each row counts pixels in a class and each column counts pixels classified as a class. By comparison of bold numbers (misclassification) in the confusion matrices, we observe that SDT algorithm reduces over one-fifth false wetland pixels and over ten percents of false dryland pixels.

Corresponding confusion maps are in Fig. 4.12 where each color represents one cell of a confusion matrix. For example, the blue color in Fig. 4.12a corresponds to the cell with value 47, 173 in the Table 4.2. The areas circled in Fig. 4.12b, c illustrate where SDT algorithm reduces false wetland pixels (in blue color) and false dryland pixels (in black color).

4.4.3 Does Incorporating Spatial Autocorrelation Reduce Salt-and-Pepper Noise?

We evaluated the salt-and-pepper level by the BB join count [14] on wetland class. $JC = \sum_{i,j} w_{i,j} x_i x_j$ where w is a w-matrix and x is one only when the pixel is a wetland class. The join count measure is then normalized. The higher the normalized joint count, the more autocorrelated the pixels are, thus the less salt-and-pepper noise there is. From Table 4.4, we see the amount of salt-and-pepper noise is reduced in the spatial decision tree learning algorithms. However, as shown in Fig. 4.12b, c, significant salt-and-pepper noise errors still remain in the spatial decision tree predictions. The reason may be that the spatial information gain-based spatial decision tree only changes tree node feature selection, without modifying the pool of candidate features. If all the candidate features tend to produce salt-and-pepper noise, the new heuristic still needs to pick one among them, impacting spatial autocorrelation of the predicted map.

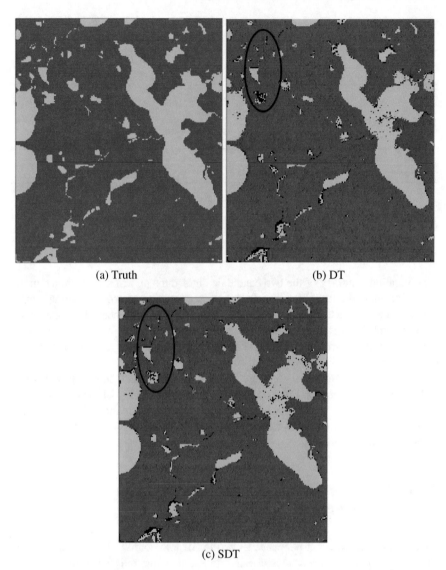

Fig. 4.12 Confusion maps and ground truth: *green* for true wetland, *red* for true dryland, *blue* for false wetland, *black* for false dryland (best viewed in color)

4.4.4   How May One Choose α, the Balancing Parameter for SIG Interestingness Measure?

This subsection describes how to automatically choose the value of α, the most important parameter in the SDT learning algorithm. The approach is to learn α via

Table 4.4 BB Join Count

	DT	SDT
Observed JC	404,454	402,586
Expected μ	102,158	95,938.2
Variance σ^2	1.24e+08	1.16e+08
$(JC - \mu)/\sigma$	**27.18**	**28.45**

Fig. 4.13 Learning balancing parameter α

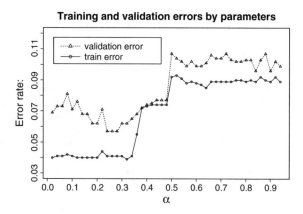

the independent validation set drawn from training area. More specifically, we first fix the other parameters as described in parameter setting part. Then, under each α value in a candidate set (i.e., 0.02, 0.04, 0.06, to 1.00), a spatial decision tree was learned from the training samples and then evaluated on validation samples. Validation errors with different α values were plotted in a curve, as indicated by Fig. 4.13. Then, the α value with minimum validation error, i.e., $\alpha = 0.26$, was selected. This method is also applicable to other study areas.

4.5 Summary

This chapter proposes a spatial decision tree learning algorithm for geographical classification problem. The problem is important in many applications but is challenging due to the presence of spatial autocorrelation. Results of a case study show that the proposed spatial decision tree learning algorithm outperforms traditional decision tree in classification accuracy and salt-and-pepper noise. We also observe that incorporating spatial autocorrelation into information gain heuristics may not be able to avoid or remove all salt-and-pepper noise in final predicted class image. The reason may be that candidate tree models are not able to explicitly control the prediction of nearby locations. The approach may be more appropriate for raster maps with a relatively smaller number of cells (e.g., state map, county map). In order to guarantee that nearby locations are predicted into similar classes, tree node tests must be extended to add spatial constraints.

Future work can be pursued to improve the method. First, we can test the sensitivity of other parameters and conduct accuracy evaluation on multiple datasets. We may also compare our spatial decision tree learning algorithm with other relevant techniques in geographical classification, e.g., spatial predictive clustering tree [15], GEOBIA (geographical object-based image analysis) [16], preprocessing, and post-processing.

References

1. M.A. Friedl, C.E. Brodley, Decision tree classification of land cover from remotely sensed data. Remote Sens. Environ. **61**(3), 399–409 (1997)
2. A. Akselrod-Ballin, M. Galun, R. Basri, A. Brandt, M. Gomori, M. Filippi, P. Valsasina, An integrated segmentation and classification approach applied to multiple sclerosis analysis, in *2006 IEEE Computer Society Conference on Computer Vision and Pattern Recognition*, vol. 1 (IEEE, 2006), pp. 1122–1129
3. M. Celebi, H. Kingravi, Y. Aslandogan, W. Stoecker, Detection of blue-white veil areas in dermoscopy images using machine learning techniques, in *Proceedings of SPIE Vol*, vol. 6144 (Citeseer, 2006), pp. 61445T–1
4. R. Brooks, D. Wardrop, C. Cole, Inventorying and monitoring wetland condition and restoration potential on a watershed basis with examples from spring creek watershed, pennsylvania, USA. Environ. Manag. **38**(4), 673–687 (2006)
5. A. Deschamps, D. Greenlee, T. Pultz, R. Saper, Geospatial data integration for applications in flood prediction and management in the red river basin, in *International Geoscience and Remote Sensing Symposium, Toronto, Canada. Symposium, Geomatics in the Era of RADARSAT (GER'97)* (Ottawa, Canada, 2002)
6. R. Hearne, Evolving water management institutions in the red river basin. Environ. Manag. **40**(6), 842–852 (2007) (Springer)
7. B. Walsh, How wetlands worsen climate change, http://www.time.com/time/health/article/0, 8599,1953751,00.html (2010)
8. J. Quinlan, Induction of decision trees. Mach. Learn. **1**(1), 81–106 (1986) (Springer)
9. J.R. Quinlan, *C4.5: Programs for Machine Learning* (Morgan Kaufmann, San Mateo, 1993)
10. B. Ripley, Classification and regression trees, in *R Package Version* (2005), p. 1
11. Z. Jiang, S. Shekhar, P. Mohan, J. Knight, J. Corcoran, Learning spatial decision tree for geographical classification: a summary of results, in *SIGSPATIAL/GIS* (ACM, 2012), pp. 390–393
12. L. Anselin, Local indicators of spatial association–lisa. Geograph. Anal. **27**(2), 93–115 (1995)
13. X. Li, C. Claramunt, A spatial Entropy-Based decision tree for classification of geographical information, in *Transactions in GIS*, vol. 10(3) (Blackwell Publishing Ltd, 2006), pp. 451–467
14. O. Schabenberger, C. Gotway, in *Statistical Methods for Spatial Data Analysis*, vol. 64 (CRC Press, 2005)
15. D. Stojanova, M. Ceci, A. Appice, D. Malerba, S. Džeroski, Global and local spatial autocorrelation in predictive clustering trees, in *Discovery Science* (Springer, Berlin, 2011), pp. 307–322
16. G. Hay, G. Castilla, Geographic object-based image analysis (GEOBIA): a new name for a new discipline, in *Object-Based Image Analysis* (Springer, Berlin, 2008), pp. 75–89

Chapter 5
Focal-Test-Based Spatial Decision Tree

Abstract This chapter introduces another spatial classification technique called focal-test-based spatial decision tree (FTSDT), in which the tree traversal direction of a sample is based on both local information and focal (neighborhood) information. We also provide comparisons of FTSDT with existing decision trees and spatial decision trees on real-world wetland mapping data.

5.1 Introduction

Given a spatial raster framework, as well as training and test sets, the spatial decision tree learning (SDTL) problem aims to find a decision tree model that minimizes classification errors as well as salt-and-pepper noise. Figure 5.1 is a motivation example from a real-world wetland mapping application. Input features are bands of three aerial photographs (Fig. 5.1a–c). Classification results by two existing decision tree classifiers [1, 2] are shown in Fig. 5.1e, f, respectively. Both predicted maps exhibit poor appearance accuracy with high levels of salt-and-pepper noise, when compared with ground truth classes (Fig. 5.1d).

Societal Applications: The SDTL problem has many applications. In the field of remote sensing, a large amount of images of the earth surface are collected (e.g., NASA collects about 5TB data per day). SDTL can be used to classify remote sensing images into different land cover types [3]. For example, in wetland mapping [4, 5], explanatory features, including spectral bands (e.g., red, green, blue, near-infrared) from remote sensors, are used to map land surface into wetland areas and dryland areas. Land cover classification is important for climate change research [6], natural resource management [7, 8], and disaster management [9]. In medical image processing, SDTL can help in lesion classification and brain tissue segmentation [10, 11] on MRI images. It can also be used for galaxy classification [12] in astronomy and semiconductor inspection [13] in materials science.

Challenges: A key challenge in the SDTL problem is that learning samples show spatial autocorrelation in class labels. For example, the ground truth class labels in Fig. 5.1d show strong spatial autocorrelation due to the phenomenon of "patches" (i.e., regions of the same class tend to be contiguous). Testing only local feature information in decision nodes results in salt-and-pepper noise, i.e., locations or pix-

© Springer International Publishing AG 2017
Z. Jiang and S. Shekhar, *Spatial Big Data Science*,
DOI 10.1007/978-3-319-60195-3_5

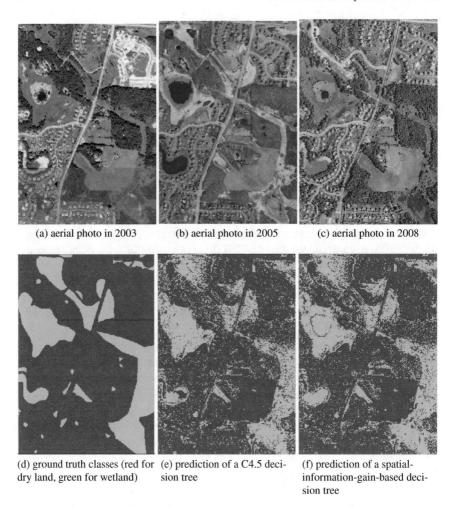

(a) aerial photo in 2003   (b) aerial photo in 2005   (c) aerial photo in 2008

(d) ground truth classes (red for   (e) prediction of a C4.5 deci-   (f) prediction of a spatial-
dry land, green for wetland)   sion tree   information-gain-based deci-
                                                                    sion tree

Fig. 5.1 Real-world problem example

els whose class labels are different from those of their neighbors, as illustrated in
Fig. 5.1e. However, incorporating focal (i.e., neighborhood) information increases
both the number and the complexity of candidate tree node tests. Instead of simple linear scanning and thresholding on one-dimensional feature values, tree node
tests must incorporate the spatial relationships of various neighborhood sizes. Thus,
SDTL problem is also computationally challenging.

Related work and limitations: Figure 5.2 presents a classification of related work.
Traditional decision tree algorithms include ID3 [14], C4.5 [1], and CART [15]).
These classifiers follow the classic assumption that learning samples are independently and identically distributed. This assumption does not hold for spatial data and
leads to salt-and-pepper noise in predictions. A second category is the spatial entropy
or information gain-based decision tree classifiers [2, 16–18]. These newer methods

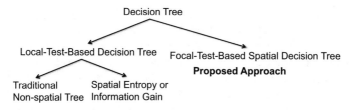

Fig. 5.2 Related work classification

use spatial autocorrelation level as well as information gain to select candidate tree node tests. While they do a better job if there exists some feature that favors spatial autocorrelation but does not provide the largest information gain in one tree node test, they still relies on local testing of information by tree nodes. Thus, if all the candidate tests have poor spatial autocorrelation, this type of decision tree will still select one of them, resulting in salt-and-pepper noise. This means neither approach adequately accounts for spatial autocorrelation in the prediction phase.

To address this limitation, we recently defined a focal-test-based spatial decision tree (FTSDT) model [19, 20], whereby the tree traversal direction of a learning sample is based on not only local but also focal (neighborhood) properties of features. We proposed FTSDT learning algorithms and evaluated the classification performance of the proposed approach on real-world remote sensing datasets. We also extended the basic FTSDT algorithm in a journal paper [21] with the following additional contributions:

1. We add a new design decision in the FTSDT model to allow focal function computation with adaptive neighborhoods (i.e., FTSDT-adaptive). Compared with previous FTSDT with fixed neighborhoods (i.e., FTSDT-fixed), the new design decision can adjust the neighborhood shape to avoid over-smoothing in wedge-shaped areas.
2. We characterize the computational structure of the FTSDT learning algorithm and confirm that the computational bottleneck is a vast number of focal function computations. We design a refined algorithm (FTSDT-Refined) that reuses focal values across candidate thresholds and prove its correctness.
3. We also provide cost models of our previous baseline algorithm and our refined algorithm and show that the refined algorithm improves computational scalability.
4. We compare the classification performance of FTSDT-adaptive with FTSDT-fixed as well as LTDT on real-world datasets. Results show that FTSDT-adaptive improves classification accuracy of FTSDT-fixed and LTDT.
5. We also conduct experimental evaluations of computational performance on real-world datasets with various parameter settings. Experiment results show that our refined algorithm significantly reduces computational time cost.

Scope: This work focuses on incorporating focal tests inside a decision tree for raster data classification. Other classification algorithms such as Markov random field [22], spatial autoregression (SAR) model [23], logistic regression, and neural network are beyond the scope. In addition, for simplicity, this work only considers learning samples with continuous features. The case of discrete features is not addressed.

Outline: The chapter is organized as follows: Sect. 5.1 introduces basic concepts and formalizes the SDTL problem; Sect. 5.2 presents our FTSDT learning algorithm, especially a new design decision to allow focal function with adaptive neighborhoods. Section 5.3 describes computational optimization and the refined algorithm design with theoretical analysis. Computational and classification performances of the proposed algorithms are evaluated in Sect. 5.4. Section 5.6 discusses some other relevant techniques in the literature. Section 5.7 concludes the chapter with future work.

5.2   Basic Concepts and Problem Formulation

This section introduces basic concepts and formally defines the spatial decision tree learning problem.

5.2.1   Basic Concepts

Spatial raster framework: A spatial raster framework F is a tessellation of a 2-D plane into a regular grid. On a spatial raster framework, there may exist a set of explanatory feature maps, as well as a class label map. For example, Fig. 5.3 shows a spatial raster framework with explanatory features $f^1, f^2, ..., f^m$ and a class label map c. Each grid cell on the raster framework is a *spatial data sample* (e.g., location i in Fig. 5.3). For simplicity, we use the words "sample," "pixel," "location," and "spatial data sample" interchangeably in the remainder of the chapter.

Neighborhood relationship: A spatial neighborhood relationship describes the range of dependency between spatial locations. It is commonly represented as a W-matrix, whose element $W_{i,j}$ has a nonzero value when locations i and j are *neighbors* and a zero value otherwise. For example, in Fig. 5.3, the pixel in dark gray has eight neighbors indicated in light gray in a 3-by-3 neighborhood.

Salt-and-pepper noise: Salt-and-pepper noise is defined as a kind of fat-tail impulse noise whose values are often extreme (e.g., minimum or maximum) [24]. In a predicted class label map, salt-and-pepper noise can be considered as a single

Fig. 5.3 Example of a spatial raster framework and a neighborhood relationship

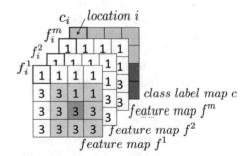

Fig. 5.4 Comparison of a
local test versus a focal test,
a local-test-based decision
tree versus a focal-test-based
spatial decision tree. ("T" is
"true"; "F" is "false")

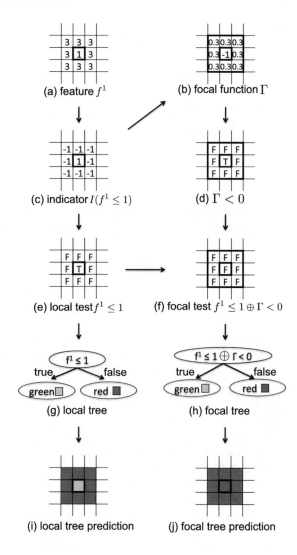

(a) feature f^1

(b) focal function Γ

(c) indicator $I(f^1 \leq 1)$

(d) $\Gamma < 0$

(e) local test $f^1 \leq 1$

(f) focal test $f^1 \leq 1 \oplus \Gamma < 0$

(g) local tree

(h) focal tree

(i) local tree prediction

(j) focal tree prediction

pixel (or a small group of contiguous pixels) that is distinct from its (or their) spatial
neighborhood. For example, in Fig. 5.4i, the central pixel is salt-and-pepper noise.

Local test and *indicators*: A *local test* $f^m \leq \delta$ checks the value of feature f^m at
a sample's location against a threshold δ. The local test results can be represented
as *indicator variable* $I(f^m \leq \delta)$ or simply I, whose value is 1 when $f^m \leq \delta$ is true
and -1 otherwise. A decision tree whose tree nodes conduct local tests is called a
<>*local-test-based decision tree (LTDT)*. For example, given the feature f^1 shown
in Fig. 5.4a, the local test results of $f^1 \leq 1$ and corresponding indicator variables are
shown in Fig. 5.4c, e, respectively. The corresponding LTDT and its class predictions
with salt-and-pepper noise are shown in Fig. 5.4g, i.

Table 5.1 List of symbols and descriptions

Symbols	Descriptions
δ	A local test threshold
$\oplus$	Logic operator "xor," i.e., $0 \oplus 1 = 1, 1 \oplus 0 = 1, 0 \oplus 0 = 0, 1 \oplus 1 = 0$
$W_{i,j}$	Neighborhood relationship between location i and location j
f^m, f_i^m	Value of feature m, the feature value at location i
$I(f^m \leq \delta), I_i$	Indicator variable of local test $f^m \leq \delta$, the indicator variable at location i
Γ, Γ_i	Focal Gamma autocorrelation statistic, the focal Gamma at location i

Focal function and *spatial autocorrelation statistic*: A focal function is an aggregate of non-spatial attribute values in the neighborhood of a location. One important kind of focal function is *focal autocorrelation statistic*, which measures the dependency between attribute values of a location and the values of its neighbors. For example, the focal Gamma index [25] on local test indicators is defined as (Table 5.1)

$$\Gamma_i = \frac{\sum_j W_{i,j} I_i I_j}{\sum_j W_{i,j}},$$

where i and j are locations, $W_{i,j}$ is a W-matrix element, and I_i and I_j are indicator variables of a local test. A negative focal Gamma value (i.e., $\Gamma < 0$) indicates that the current location is potentially salt-and-pepper noise. Figure 5.4b shows an example of focal Gamma values computed on indicator variables in Fig. 5.4c with a 3-by-3 neighborhood. The central location has a negative Gamma because its local test result is different from its neighbors'.

Focal test: A focal test is a test or a combination of tests on attribute values in a neighborhood of a location. For example, $f \leq \delta \oplus \Gamma < 0$, where $\oplus$ is an "xor" logical operator, is a focal test that combines a local test $f \leq \delta$ and the test $\Gamma < 0$. This combined focal test is less prone to salt-and-pepper noise, compared with the local test $f \leq \delta$ only. The reason is that salt-and-pepper noise pixels often have a negative focal Gamma index (i.e., $\Gamma < 0$ is true), and their local test results ($f \leq \delta$) are flipped by logical operator $\oplus$ (i.e., "false" xor true becomes "true," and "true" xor true becomes "false"). For instance, the local test result of the central pixel in Fig. 5.4e is true, but false for its neighborhood; while the focal test result of the same pixel in Fig. 5.4f is false, the same as for its neighborhood.

Focal-test-based spatial decision tree (FTSDT): An FTSDT is a tree whose nodes conduct focal tests. An example of FTSDT is in Fig. 5.4h, and its class predictions are in Fig. 5.4j. In our approach, both local tests and focal tests are defined on a single feature. When multiple features exist, the local test or focal test on each feature is considered as a candidate tree node test and the best candidate test is selected for a tree node, similar to the situation of a traditional decision tree.

5.2.2 Problem Definition

Based on the concepts above, the spatial decision tree learning problem is formally defined as follows:

Given:
- A spatial raster framework F
- A spatial neighborhood definition and its maximum size S_{max}
- Training and test samples drawn from F

Find:
- A decision tree model based on training samples.

Objective:
- Minimize classification errors as well as salt-and-pepper noise

Constraints:
- Training samples form contiguous patches of locations in F
- Spatial autocorrelation exists in class labels

Problem description: The output decision tree model can be a local-test-based decision tree or a focaltest-based spatial decision tree, depending on the selected approach. Parameters to be learned from the training set are the tree structure, as well as which feature f, test thresholds δ, and proper neighborhood size s (in the case of FTSDT) to use in each tree node.

Example: Consider the example of problem inputs and outputs in Fig. 5.5. The raster spatial framework F, shown in Fig. 5.5a, consists of training pixels on the upper half and test pixels on the lower half. Neighborhood relationship is defined as a 3-by-3 square window. The minimum tree node size is four. Figure 5.5b shows a candidate feature f^1. Figure 5.5c shows the ground truth class labels. The output local-test-based traditional decision tree learned from the training set and its predictions with salt-and-pepper noise are shown in Fig. 5.5d. In contrast, the output focal-test-based spatial decision tree and its predictions without salt-and-pepper noise are shown in Fig. 5.5e.

5.3 FTSDT Learning Algorithms

This section describes the baseline FTSDT learning algorithm (i.e., without computational optimization) of the focal-test-based spatial decision tree. The learning algorithm has two phases: a training phase, FTSDT-Train; and a prediction phase, FTSDT-Predict. FTSDT-Train here extends the previous one we proposed in [19] by allowing focal function tests with adaptive neighborhoods to avoid over-smoothing in wedge-shaped areas.

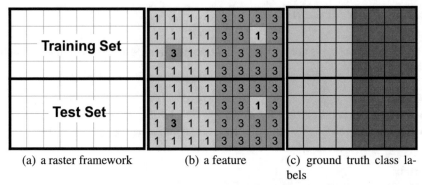

(a) a raster framework   (b) a feature   (c) ground truth class labels

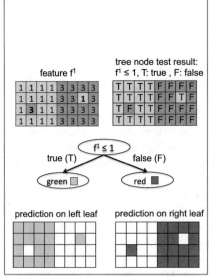

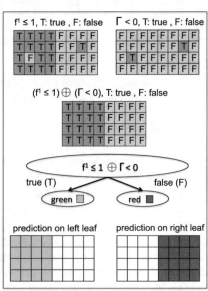

(d) an LTDT and its predictions   (e) an FTSDT and its predictions

Fig. 5.5 Illustrative problem example (best viewed in color)

5.3.1 Training Phase

FTSDT-Train (Algorithm 1) learns an FTSDT classifier from training samples. It includes two subroutines (Node-Split and Focal Function). Similar to traditional C4.5 [1], it is a divide and conquer method with a greedy strategy (i.e., maximize information gain).

Steps 1–3 check the stopping criteria. If the training samples are less than the minimum tree node size, or all the class labels are identical, a leaf labeled with the majority class will be returned.

Steps 4–13 enumerate through every candidate feature f, every neighborhood size s, and every candidate threshold δ to select the best setting for a model tree node. Candidate thresholds δ are generated from distinct values of feature f in the training samples (steps 8–9). Step 10 calls a Node-Split subroutine to split training samples. Step 11 evaluates the corresponding information gain on the column of class labels. Steps 12–13 update the current best candidate test.

Steps 14–18 create an internal node with the best test, split the training samples into two subsets accordingly, recursively call FTSDT-Train on each subset, and return the internal node.

Algorithm 2 FTSDT-Train(T, S_{max}, N_0, *neighType*)

Input:
- T: rows are samples, columns are features (last column as class)
- S_{max}: maximum neighborhood size
- N_0: minimum decision tree node size
- *neighType*: neighborhood type, 0 for fixed neighborhood, 1 for adaptive neighborhood

Output:
- root of an FTSDT model

1: let N be number of samples, F be number of features, c be column index of classes; $IG_0 = -\infty$

2: **if** $N < N_0$ or T same class **then**
3:    **return** a leaf node;
4: **for each** $f \in \{1...F\}$ **do**
5:    sort rows of T //in ascending order of fth column
6:    **for each** $s \in \{0...S_{max}\}$ **do**
7:      **for each** $i \in \{N_0...(N - N_0)\}$ **do**
8:        **if** $T[i][f] < T[i + 1][f]$ **then**
9:          $\delta = T[i][f]$
10:          $\{T_1, T_2\} = $ **Node-Split**(T, f, δ, s, *neighType*);
11:          $IG = $ **InformationGain**($T[\ ][c]$, $T_1[\ ][c]$, $T_2[\ ][c]$)
12:          **if** $IG > IG_0$ **then**
13:            $IG_0 = IG$; $s_0 = s$; $f_0 = f$; $\delta_0 = \delta = T[i][f]$;
14: $I = $ CreateInternalNode(f_0, δ_0, s_0);
15: $\{T_1, T_2\} = $ **Node-Split**(T, f_0, δ_0, s_0, *neighType*);
16: $I.left = $ **FTSDT-Train**(T_1, S_{max}, N_0, *neighType*)
17: $I.right = $ **FTSDT-Train**(T_2, S_{max}, N_0, *neighType*)
18: **return** I

Node-Split: The Node-Split subroutine (Algorithm 3) splits the training samples into two subsets based on their focal test results and proceeds as follows:

Step 1 initializes the two subsets as empty sets. Samples with node test results TRUE will be assigned to one subset, and samples with test results FALSE will be assigned to the other.

Steps 2–11 compute the focal tree node test result of each training sample and add the sample to its appropriate subset accordingly. The algorithm begins by computing local test indicators (I) of all samples. It then computes the focal function value (*focalFun[i]*) via a *FocalFunction* subroutine and computes the focal test result (*focalTest*[*i*]) on each sample location. For example, we may specify the focal function as Γ, and the focal test as "$f \leq \delta \oplus \Gamma < 0$".

Algorithm 3 Node-Split(*T*, *f*, δ, *S*, *neighType*)

Input:
- *T*: rows as samples, columns as features (last column as class)
- *f*: a feature index
- δ: threshold of feature test
- *S*: neighborhood size
- *neighType*: neighborhood type, 0 for fixed neighborhood, 1 for adaptive neighborhood

Output:
- $\{T_1, T_2\}$: sample subsets with test results true and false respectively

1: $T_1 = T_2 = \emptyset$
2: **for each** $i \in \{1...N\}$ **do**
3:     indicators $I[i] = I(T[i][f] \leq \delta)$
4: **for each** $i \in \{1...N\}$ **do**
5:     $focalFun[i] = $ **FocalFunction**$(I[], i, s, neighType)$
6:     $focalTest[i] = FocalTest(I[i], focalFun[i])$
7:     **if** $focalTest[i] == true$ **then**
8:         $T_1 = T_1 \cup \{T[i]\}$
9:     **else**
10:        $T_2 = T_2 \cup \{T[i]\}$
11: **return** $\{T_1, T_2\}$

Algorithm 4 FocalFunction(*I*, *i*, *s*, *neighType*)

Input:
- *I*: vector of indicator variable values
- *i*: current location
- *s*: neighborhood window size
- *neighType*: neighborhood type, 0 for fixed neighborhood, 1 for adaptive neighborhood

Output:
- *FocalFun*[*i*]: focal function value at current location *i*

1: identify the 2s+1 by 2s+1 window centered on location *i*
2: **if** $adaptNeigh == 0$ **then**
3:     $W_{i,j} = 1$ for all *j* in the window, $W_{i,j} = 0$ otherwise.
4: **else**
5:     get connected components of same *I* values in the window
6:     identify the topologically outermost component cc_0 that contains or surrounds location *i*
7:     $W_{i,j} = 1$ for all *j* in cc_0, $W_{i,j} = 0$ otherwise.
8: compute focal function value *foc* at location *i* based on $W_{i,j}$
9: **return** *foc*

FocalFunction: The FocalFunction subroutine (Algorithm 4) computes the focal function values of local test indicators in the neighborhood of a location. It has an important parameter *neighType*, whose value is 0 for a fixed neighborhood, and is 1 for an adaptive neighborhood. The intuition behind an adaptive neighborhood is to utilize spatial topological relationships to select proper neighbors of the central pixel in a fixed window.

Step 1 identifies all locations within the window of size $2s + 1$ by $2s + 1$ centered on the current location. These locations are potential neighbors of i.

Steps 2 and 3 determine that all the locations in the window are neighbors of i if a fixed neighborhood is used, similar to our previous work in [19].

Steps 4–7 determine which locations in the window are neighbors if an adaptive neighborhood is used. The window is first segmented into different connected components, each of which has the same indicator value. Then the component that is the outermost, and that surrounds or contains the current location i, is considered as the actual set of neighbors.

Steps 8 and 9 compute a focal function value based on the neighbors identified and return the value.

Illustration: The entire execution trace of FTSDT-Train with fixed neighborhoods can be found in [19]. Due to space limitations, here we only illustrate the extension of focal tests with adaptive neighborhoods by comparing them with fixed neighborhoods. Consider the example in Fig. 5.6, which describes one iteration of candidate test selection (steps 9 to 10 in Algorithm 2). Assume that the current neighborhood window size is $s = 2$ (i.e., 5-by-5).

Figure 5.6a–c shows current candidate feature f, ground truth classes, and local test indicators on the current test threshold $\delta = 1$, respectively. The feature f map (Fig. 5.6a) contains a wedge-shaped patch (fifteen pixels with feature value 1 on the lower left corner) and three salt-and-pepper noise pixels. The pixels on the wedge-shaped patch are not salt-and-pepper noise and thus should not be smoothed (i.e., should avoid over-smoothing).

Figure 5.6d–f shows the focal test results with fixed neighborhoods. For instance, Fig. 5.6d highlights the fixed neighborhood (in light gray) of a central pixel (in dark gray), which contains too many irrelevant pixels (with indicator value -1) outside the wedge-shaped patch (the one we previously describe in the last paragraph). The focal function Γ of the central pixel is -0.3 (i.e., $\Gamma < 0$) as shown in dark gray in Fig. 5.6e, mistakenly indicating that it is salt-and-pepper noise. Thus, its final focal test result mistakenly flips its local test result from "true" to "false". Similarly, several other pixels in the wedge-shaped patch are also over-smoothed. The final over-smoothed focal test results of the patch are shown in dark gray in Fig. 5.6f.

In contrast, the bottom row of Fig. 5.6 shows focal test results with adaptive neighborhoods. Figure 5.6g highlights an adaptive neighborhood (in light gray) of the central pixel (in dark gray), which is a connected component (with indicator value 1) that contains the central dark pixel. The focal function Γ of the central pixel is now 1 (Fig. 5.6b) based on the adaptive neighborhood. Thus, the final focal test is still "true" ($\Gamma < 0$ is false, "true" xor false is still "true"). The three salt-and-pepper noise pixels are still smoothed. Comparing Fig. 5.6f and i, it is clear that focal tests with adaptive neighborhoods can better separate the two classes (i.e., give higher information gain) due to less over-smoothing of the wedge-shaped area.

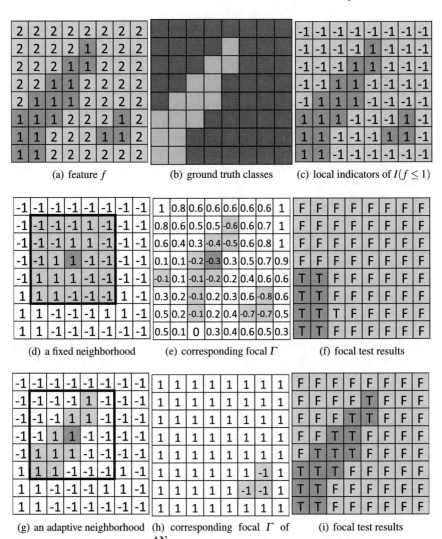

(a) feature f (b) ground truth classes   (c) local indicators of $I(f \le 1)$

(d) a fixed neighborhood   (e) corresponding focal Γ (f) focal test results

(g) an adaptive neighborhood   (h) corresponding focal Γ of AN   (i) focal test results

Fig. 5.6 Comparison of focal tests with fixed and adaptive neighborhoods, $s = 2$ (i.e., 5 by 5 window)

5.3.2   Prediction Phase

The FTSDT-Predict algorithm (Algorithm 5) uses an FTSDT to predict the class labels of test samples based on their feature values and a spatial neighborhood structure. It is a recursive algorithm. If the tree node is a leaf, then the class label of the leaf is assigned to all current samples. Otherwise, samples are split into two subsets according to the focal test results in the root node, and each subset is classified by its corresponding subtree.

Algorithm 5 FTSDT-Predict(R, T)

Input:
- R: root of an FTSDT model
- T: rows as samples, columns as features

Output:
- C: $C[i]$ as class label of ith sample

1: **if** $R.type == Leaf$ **then**
2:     assign C with $R.class$
3:     **return** C
4: $f_0 = R.f, \delta_0 = R.\delta, s_0 = R.s$
5: $\{T_1, T_2\} = $ **Node-Split** (T, f_0, δ_0, s_0)
6: $C_1 = $ **FTSDT-Predict** $(R.Left, T_1)$;
7: $C_2 = $ **FTSDT-Predict** $(R.Right, T_2)$;
8: **return** $C = $ combine(C_1, C_2)

5.4 Computational Optimization: A Refined Algorithm

This section addresses the computational challenges of the focal-test-based spatial decision tree learning process. It first identifies the computational bottleneck of the baseline training algorithm; proposes a refined algorithm; proves its correctness; and finally provides a cost model for the computational complexity. For simplicity, examples in this section are with a fixed neighborhood. However, the proposed refined algorithm and its analysis are also applicable to the case of adaptive neighborhoods.

5.4.1 Computational Bottleneck Analysis

Recall that the baseline algorithm (Algorithm 2) calls a Node-Split subroutine for every distinct value (i.e., candidate threshold) on every feature and neighborhood size. Each call involves focal function computations for all samples and is, therefore, a likely computational bottleneck. To verify this hypothesis, we conducted computational bottleneck analysis with parameter settings $S_{max} = 5$ and $N_0 = 50$. The results, shown in Fig. 5.7, confirm that the focal function computation accounts for the vast majority of total time cost. Furthermore, this cost increases much faster than other costs as training sample sizes increase.

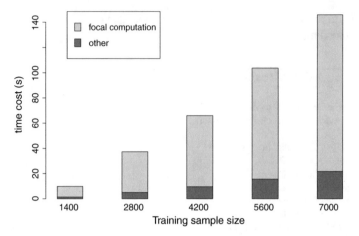

Fig. 5.7 Computational bottleneck analysis in training algorithms

5.4.2   A Refined Algorithm

To reduce the computational bottleneck shown above, we designed a refined approach
called *cross-threshold-reuse*. This approach is based on the observation that when the
candidate threshold value increases, only a small number of samples have their local
and focal test results updated. In other words, once computation is completed for one
candidate threshold, the test results of most samples will remain the same and can
be reused for consecutive thresholds. An illustrative example is given in Fig. 5.8a–c,
where the values of feature f are shown in Fig. 5.8a, and the local indicators and
focal gamma values for test thresholds $\delta = 1$ and $\delta = 2$ are shown in Fig. 5.8b, c,
respectively. As can be seen, only location 2 and its neighbors update local indicators
and focal values (shown in gray color in Fig. 5.8b).

The cross-threshold-reuse approach updates one sample at a time together with
its neighbors. The details of this approach are given in the algorithm FTSDT-Train-
Refined (Algorithm 6). The key difference from previous FTSDT-Train is that the
refined algorithm calls the Node-Split subroutine (Algorithm 3) only once. For sub-
sequent sample indices (potential candidate thresholds), it calls Node-Split-Update
subroutine (Algorithm 7) instead. More specifically, in step 7, it finds the first effective candidate threshold (guaranteed minimum node size N_0). In step 8, it enumerates
all possible sample indices. If it is the first enumeration, Node-Split is called to completely compute and memorize local and focal results for all samples (steps 9 to 10).
Otherwise, Node-Split-Update is called to avoid redundant computation (step 12).
Steps 13–16 check if the current split index i is an effective split threshold, and if so,
information gain is evaluated to maintain the current best candidate test.

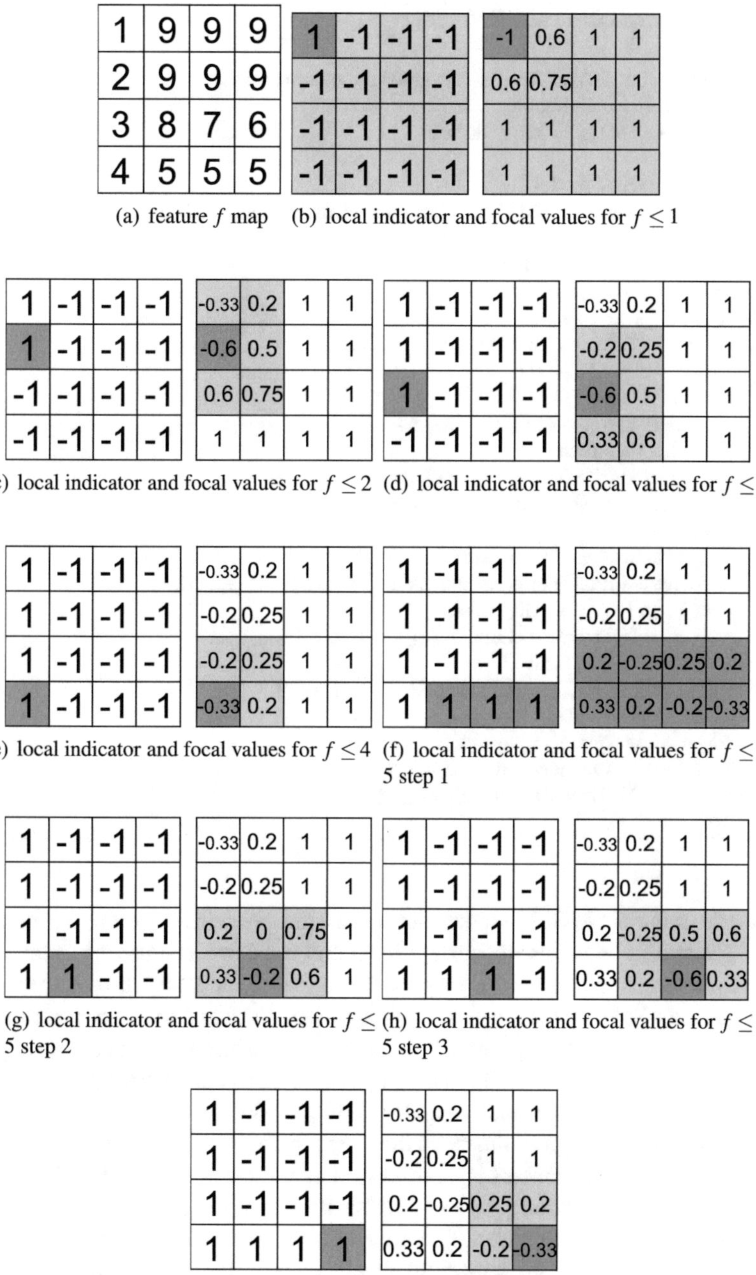

1	9	9	9
2	9	9	9
3	8	7	6
4	5	5	5

1	-1	-1	-1
-1	-1	-1	-1
-1	-1	-1	-1
-1	-1	-1	-1

-1	0.6	1	1
0.6	0.75	1	1
1	1	1	1
1	1	1	1

(a) feature f map   (b) local indicator and focal values for $f \leq 1$

1	-1	-1	-1
1	-1	-1	-1
-1	-1	-1	-1
-1	-1	-1	-1

-0.33	0.2	1	1
-0.6	0.5	1	1
0.6	0.75	1	1
1	1	1	1

1	-1	-1	-1
1	-1	-1	-1
1	-1	-1	-1
-1	-1	-1	-1

-0.33	0.2	1	1
-0.2	0.25	1	1
-0.6	0.5	1	1
0.33	0.6	1	1

(c) local indicator and focal values for $f \leq 2$ (d) local indicator and focal values for $f \leq 3$

1	-1	-1	-1
1	-1	-1	-1
1	-1	-1	-1
1	-1	-1	-1

-0.33	0.2	1	1
-0.2	0.25	1	1
-0.2	0.25	1	1
-0.33	0.2	1	1

1	-1	-1	-1
1	-1	-1	-1
1	-1	-1	-1
1	1	1	1

-0.33	0.2	1	1
-0.2	0.25	1	1
0.2	-0.25	0.25	0.2
0.33	0.2	-0.2	-0.33

(e) local indicator and focal values for $f \leq 4$ (f) local indicator and focal values for $f \leq$ 5 step 1

1	-1	-1	-1
1	-1	-1	-1
1	-1	-1	-1
1	1	-1	-1

-0.33	0.2	1	1
-0.2	0.25	1	1
0.2	0	0.75	1
0.33	-0.2	0.6	1

1	-1	-1	-1
1	-1	-1	-1
1	-1	-1	-1
1	1	1	-1

-0.33	0.2	1	1
-0.2	0.25	1	1
0.2	-0.25	0.5	0.6
0.33	0.2	-0.6	0.33

(g) local indicator and focal values for $f \leq$ (h) local indicator and focal values for $f \leq$
5 step 2   5 step 3

1	-1	-1	-1
1	-1	-1	-1
1	-1	-1	-1
1	1	1	1

-0.33	0.2	1	1
-0.2	0.25	1	1
0.2	-0.25	0.25	0.2
0.33	0.2	-0.2	-0.33

(i) local indicator and focal values for $f \leq$
5 final

Fig. 5.8 Illustrative example of redundant focal Γ computation

---

Algorithm 6 FTSDT-Train-Refined(T, S_{max}, N_0)

Input:
- T: rows as samples, columns as features (last column as class)
- S_{max}: maximum neighborhood size
- N_0: minimum decision tree node size

Output:
- root of an FTSDT model

1: denote N as number of samples , F as number of features, c as column index of classes ;
   $IG_0 = -\infty$
2: **if** $N < N_0$ or T same class **then**
3:    **return** a leaf node;
4: **for each** $f \in \{1...F\}$ **do**
5:    sort rows of T //in ascending order of fth column
6:    **for each** $s \in \{0...S_{max}\}$ **do**
7:       i_0 = first i with $T[i][f] > T[N_0][f]$
8:       **for each** $i \in \{(i_0 - 1)...(N - N_0)\}$ **do**
9:          **if** first time **then**
10:             // memorize $indicator, focalFunc, T_1, T_2$
               $\{T_1, T_2\} = $ **Node-Split**$(T, f, \delta = T[i][f], s)$
11:          **else**
12:             // update $indicator, focalFunc, T_1, T_2$
               Node-Split-Update($indicator, focalFunc, i, s,$ $\{T_1, T_2\}$)
13:          **if** $T[i][f] < T[i + 1][f]$ **then**
14:             $IG = $ **InformationGain**($T[\ ][c], T_1[\ ][c], T_2[\ ][c]$)
15:             **if** $IG > IG_0$ **then**
16:                $IG_0 = IG; s_0 = s; f_0 = f; \delta_0 = \delta = T[i][f]$
17: $I = $ CreateInternalNode(f_0, δ_0, s_0);
18: $\{T_1, T_2\} = $ **Node-Split**(T, f_0, δ_0, s_0);
19: $I.left = $ **FTSDT-Train-Refined**(T_1, S_{max}, N_0)
20: $I.right = $ **FTSDT-Train-Refined**(T_2, S_{max}, N_0)
21: **return** I

Details of the Node-Split-Update subroutine are given in Algorithm 7. The input of this subroutine includes the current local and focal test results, the index of the currently enumerated sample i, neighborhood size s, and the current split result $\{T_1, T_2\}$. Node-Split-Update begins by updating the local and focal tests of sample i, and adjusting $\{T_1, T_2\}$ accordingly. Then, it updates the test results of every neighbor j of sample i and adjusts $\{T_1, T_2\}$. These updates all carry a constant time cost since these are done in only one small neighborhood window.

Execution trace: Fig. 5.8 illustrates the execution trace of steps 8 to 12 of the new update algorithm. The context is as follows: Feature f is shown in Fig. 5.8a; $s = 1$ (3-by-3 fixed neighborhood); $N_0 = 1$; the local indicator is $I(f \leq \delta)$; and the focal function is Γ as before. Figure 5.8b–i are local indicators and focal function values under different candidate test thresholds (1 to 5). The refined algorithm only updates the local indicators and focal values, shown in gray colors of Fig. 5.8c–e and g–i.

Algorithm 7 Node-Split-Update(*indicator*, *focalFunc*, *i*, *s*, $\{T_1, T_2\}$)

Input:
- *indicator*: array of local result as $I(f \leq \delta)$
- *focalFunc*: array of focal function values, e.g., Γ_I^S
- *i*: index of sample shifted below threshold
- *s*: neighborhood size
- $\{T_1, T_2\}$: two subsets of samples

1: *indicator*[*i*] = 1; update *focalFunc*[*i*], and then $\{T_1, T_2\}$
2: **for each** $j \in N^s(i)$ // $N^s(i)$ is *i*'s neighborhood of size *s* **do**
3:    update *indicator*[*j*], *focalFunc*[*j*], and then $\{T_1, T_2\}$

5.4.3 Theoretical Analysis

We now prove the correctness of proposed computationally refined algorithm. We also provide a cost model of computational complexity. The proof of correctness is non-trivial, because when the candidate threshold changes, multiple sample locations as well as their neighbors may need to update their focal values (e.g., Fig. 5.8f), and these updates are at the same time. However, our approach still simply changing one sample location as well as its neighbors each time (e.g., Fig. 5.8g–i), and it has the same results.

Theorem 5.1 *The FTSDT-Train-Refined algorithm is correct, i.e., it returns the same output as FTSDT-Train.*

Proof We need only look at steps 7–16. The initial value of *i* in the for-loop here (i.e., one step ahead of the first sample whose feature value is greater than that of N_0) is the same as that value in FTSDT-Train, due to the if-condition in step 8 of FTSDT-Train. Now, we focus on the local and focal computation parts in steps 9–12. We will prove it in two cases as below.

Case 1: A new threshold shifts only one sample i, i.e., $T[i][f] < T[i+1][f]$. In this case, the local result $I(f \leq \delta)$ changes only on sample *i*, i.e., $I[i] = 1$, while the feature value *f* is unchanged. Since the focal function of a sample only depends on *f* and *I* in its neighborhood, its value changes only on sample *i* and its neighbor *j*. Thus, Node-Split-Update in this case is correct. One example of this case appears in Fig. 5.8b, c.

Case 2: a new threshold shifts multiple samples, i.e., $T[i][f] = T[i+1][f] = \ldots = T[i+k][f] < T[i+k+1][f]$. In this case, our refined algorithm still only updates sample *i* and its neighbors, as though $T[i][f] < T[i+1][f]$ (in other words, as though $T[i][f]$ were an effective candidate threshold). This updating process continues until new *i* becomes $i + k$, i.e., the next effective candidate threshold. If the feature value $T[i][f]$ was strictly increasing, the final local and focal values should be correct, as proved in case 1. Meanwhile, it is also obvious that whether feature values are strictly increasing or not before $i = i + k$ does not influence the final local and focal values. Thus, the final updated result for $i = i + k$ is also correct. An example for this case is given in Fig. 5.8c, d.

To analyze the cost model of the two proposed training algorithms, we denote the following variables:

- N: number of samples
- N_d: number of distinct feature values
- S_{max}: maximum neighborhood size
- N_0: minimum tree node size
- F: number of features

Lemma 5.1 *The baseline algorithm FTSDT-Train has a time complexity of* $O(FN^2(\log N + N_d S_{max}^3)/N_0)$.

Proof Given N samples and minimum node size N_0, tree node number is at most N/N_0, i.e., $O(N/N_0)$. For each tree node, the algorithm sorts samples for all features and enumerates through all $O(N_d)$ thresholds for all the F features under all the $S_{max} + 1$ different neighborhood sizes. In each enumeration, a Node-Split subroutine is called, which has time complexity $O(NS_{max}^2)$, where $O(S_{max}^2)$ is the number of neighbors under a square neighborhood. Thus, for each node, the time cost is $O(F \cdot (N \log N + S_{max} \cdot N_d \cdot NS_{max}^2)) = O(FN(\log N + N_d S_{max}^3))$. Finally, the total time cost is $O(N/N_0 \cdot FN(\log N + N_d S_{max}^3)) = O(FN^2(\log N + N_d S_{max}^3)/N_0)$.

Lemma 5.2 *The SDT-Train-Refined algorithm has a time complexity of* $O(FN^2(\log N + S_{max}^3)/N_0)$.

Proof The number of tree nodes is $O(N/N_0)$. For each node and each of the $O(F)$ features, the refined algorithm sorts and enumerates through all $O(N)$ samples under all $O(S_{max} + 1)$ neighborhood sizes. Node-Split is called only once (with time cost $O(NS_{max}^2)$) in these enumerations, and Node-Split-Update is called for the rest (each with time cost $O(S_{max}^2)$). Thus, for each node, the time cost is $O(F \cdot N \cdot \log N + FS_{max} \cdot (NS_{max}^2 + N \cdot S_{max}^2)) = O(FN(\log N + S_{max}^3))$, and the total time cost is $O(FN^2(\log N + S_{max}^3)/N_0)$.

Theorem 5.2 *FTSDT-Train-Refined is faster than FTSDT-Train when $N_d \gg 1$ (i.e., N_d is much greater than 1).*

Proof From the lemmas above, the cost models of the two algorithms only differ in one factor, which is $O(\log N + N_d S_{max}^3)$ for FTSDT-Train and $O(\log N + S_{max}^3)$ for the refined algorithm. Since $N_d \gg 1$, the cost of the refined algorithm is always smaller. The same can be proved by simplifying the two cost models, if we assume $N_d \propto N$, and F, S_{max}, and N_0 are constants. Then, the cost of FTSDT-Train is $O(N^3)$, while the cost of FTSDT-Train-Refined is $O(N^2 \log N)$, as shown in Table 5.2. Note that the condition $N_d \gg 1$ is often satisfied for continuous features.

Table 5.2 Simplified cost model with different numbers of distinct feature values (N_d)

Algorithm	$N_d = O(1)$	$N_d = O(N)$
FTSDT-Train	$O(N^2 \log N)$	$O(N^3)$
FTSDT-Train-Refined	$O(N^2 \log N)$	$O(N^2 \log N)$

5.5 Experimental Evaluation

The goal was to investigate the following questions:

- How do LTDT, FTSDT-fixed, and FTSDT-adaptive compare with each other in classification accuracy?
- How do LTDT, FTSDT-fixed, and FTSDT-adaptive compare with each other in salt-and-pepper noise level?
- Does the FTSDT-Train-Refined algorithm reduce the computational cost of baseline FTSDT-Train algorithm?

5.5.1 Experiment Setup

Experiment design: The experiment design is shown in Fig. 5.9. To evaluate classification performance, we compared the LTDT learner (i.e., C4.5), the FTSDT learner with fixed neighborhoods (i.e., FTSDT-fixed), as well as the FTSDT learner with adaptive neighborhoods (i.e., FTSDT-adaptive) on test accuracy and autocorrelation level. To evaluate computational performance, we used FTSDT with fixed neighborhoods for simplicity and compared the baseline approach (i.e., FTSDT-Train) with the computationally refined approach (i.e., FTSDT-Train-Refined). Computational time reported was the average of 10 runs. All the algorithms were implemented in C language. Experiments were conducted on a Dell workstation with Quad-core Intel Xeon CPU E5630 @ 2.53 GHz, and 12 GB RAM.

Dataset description: We used high resolution (3m-by-3m) remote sensing imagery collected from the city of Chanhassen, MN, by the National Agricultural Imagery Program and Markhurd Inc. There were 12 continuous explanatory features including multi-temporal (for the years 2003, 2005, and 2008) spectral information (R, G, B, NIR) and normalized difference vegetation index (NDVI). Class labels (wet land and dry land) were created by a field crew and photograph interpreters between 2004 and 2005.

To evaluate classification performance, we selected two scenes from the city. On each scene, we used systematic clustered sampling to select a number of wetland and dryland contiguous clusters of pixels as the training set and the remaining pixels

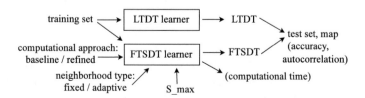

Fig. 5.9 Experiment design

Table 5.3 Description of datasets

Scene	Size	Training samples
1	476 by 396	11837(dryland class); 5679 (wetland class)
2	482 by 341	7326 (dryland class); 2735 (wetland class)

as test sets. More details are given in Table 5.3. To evaluate computational performance, we used scene 1 and created training sets with different sizes and number of distinct feature values to test sensitivity of computational cost on various settings. The variables tested were previously defined in Sect. 4.3.

Choice of focal test functions: For the focal-test-based spatial decision tree, we used the specific focal test $(f \leq \delta) \oplus (\Gamma < 0)$ described in Sect. 2.1.

5.5.2 Classification Performance

5.5.2.1 How Do LTDT, FTSDT-Fixed, and FTSDT-Adaptive Compare in Classification Accuracy?

Parameter settings were $S_{max} = 5$, $N_0 = 200$ for the first dataset, and $N_0 = 50$ for the second dataset. We compared the classification performance of the proposed FTSDT-adaptive and FTSDT-fixed with LTDT in terms of confusion matrices, precision and recall, and F-measure (i.e., harmonic mean of precision and recall) on the test set. The results are listed in Table 5.4. In the confusion matrix, the columns are test samples classified as dryland and wetland, respectively, and the two rows are test samples whose true class labels are dryland and wetland, respectively. Precision and recall were computed on the wetland class. As can be seen, on the first dataset, FTSDT-fixed improves the F-measure of LTDT from 0.78 to 0.81 (e.g., false negatives decrease by around 20% from 15346 to 12470), and FTSDT-adaptive further improves the F-measure of FTSDT-fixed from 0.81 to 0.83 (e.g., false negatives decrease by around 15% from 12470 to 10618). Similar improvements are also seen in the results on the second dataset.

We also conducted significance tests on the difference of the confusion matrices between LTDT and FTSDT-fixed, and between FTSDT-fixed and FTSDT-adaptive. The statistic used was $\hat{K}$ (estimate of Kappa coefficient) [26–28], defined as

$$\hat{K} = \frac{n \sum_{i=1}^{k} n_{ii} - \sum_{i=1}^{k} n_{i+}n_{+i}}{n^2 - \sum_{i=1}^{k} n_{i+}n_{+i}},$$

where n is the sum of all elements, and n_{ii}, n_{i+}, and n_{+i} are the diagonal, row sum, and column sum, respectively. $\hat{K}$ reflects the degree to which a confusion matrix is different from a random guess. First, we computed $\hat{K}$ and its variance for each

Table 5.4 Classification performance of LTDT, FTSDT-fixed, and FTSDT-adaptive

Scene	Models	Confusion matrix		Precision	Recall	F-score	Γ index
1	LTDT	99141	10688	0.81	0.75	**0.78**	0.87
		15346	45805				
	FTSDT-fixed	99755	10074	0.83	0.80	**0.81**	0.96
		12470	48681				
	FTSDT-adapt	99390	10439	0.83	0.83	**0.83**	0.93
		10618	50533				
2	LTDT	104615	10820	0.70	0.66	**0.68**	0.87
		13254	25612				
	FTSDT-fixed	107297	8138	0.77	0.69	**0.73**	0.96
		11984	26882				
	FTSDT-adapt	105999	9436	0.76	0.75	**0.75**	0.92
		9744	29122				

Table 5.5 $\hat{K}$ statistics of confusion matrices

Scene	LTDT		FTSDT-fixed		FTSDT-adaptive	
	$\hat{K}$	$\hat{K}$ variance	$\hat{K}$	$\hat{K}$ variance	$\hat{K}$	$\hat{K}$ variance
1	0.66	3.6×10^{-6}	0.71	3.2×10^{-6}	0.73	3.0×10^{-6}
2	0.58	5.9×10^{-6}	0.64	5.3×10^{-6}	0.67	4.8×10^{-6}

Table 5.6 Significance test on difference of confusion matrices

Scene	LTDT versus FTSDT-fixed		FTSDT-fixed versus FTSDT-adaptive	
	Z score	Result	Z score	Result
1	18.2	Significant	8.6	Significant
2	19.4	Significant	8.5	Significant

evaluation candidate as shown in Table 5.5. Then, we conducted a Z-test on pairs of $\hat{K}$ statistics of LTDT and FTSDT-fixed, as well as FTSDT-fixed and FTSDT-adaptive. The results show that improvements of FTSDT-adaptive over LTDT and FTSDT-fixed in confusion matrices are significant (Table 5.6).

5.5.2.2 How Do LTDT, FTSDT-fixed, and FTSDT-adaptive Compare with Each Other in Salt-and-pepper Noise Level?

We compared prediction maps by LTDT, FTSDT-fixed, and FTSDT-adaptive on the amount of salt-and-pepper noise, as measured by a spatial autocorrelation statistic, i.e., gamma index Γ with queen neighborhoods. This index ranges from 0 to 1, and a larger index value indicates less salt-and-pepper noise. Parameter settings were $S_{max} = 5$, $N_0 = 200$ for the first dataset, and $N_0 = 50$ for the second dataset. The last column of Table 5.4 shows the spatial autocorrelation levels of the LTDT, FTSDT-fixed, and FTSDT-adaptive predictions on the two datasets. As can be seen, both FTSDT-fixed and FTSDT-adaptive improve the spatial autocorrelation (i.e., reducing salt-and-pepper noise) over LTDT significantly. The spatial autocorrelation of FTSDT-adaptive predictions is somewhat lower than for FTSDT-fixed since it uses flexible neighborhoods to avoid FTSDT-fixed's over-smoothing. Nonetheless, the overall classification accuracy of FTSDT-adaptive is better.

5.5.2.3 Case Study

We ran a case study to illustrate the difference of predictions among the LTDT, the FTSDT-fixed, and FTSDT-adaptive learning algorithms. The dataset was again the Scene 1 images from the city of Chanhassen. Several of the input multi-temporal optical features are mapped in Fig. 5.10a, b. Target classes were wetland and dryland. The maximum neighborhood size was set to 5 (11-by-11 window), and minimum tree node size was 200.

The predictions of LTDT, FTSDT-fixed, and FTSDT-adaptive are shown in Fig. 5.10c–e, respectively. The green and red colors represent correctly classified wetland and correctly classified dryland. The black and blue colors represent errors of false wetland and false dryland. As can be seen, the prediction by LTDT has lots of salt-and-pepper noise (in black and blue colors) due to the high local variation of features within wetland or dry land patches. The predictions of FTSDT-fixed (Fig. 5.10d) and FTSDT-adaptive (Fig. 5.10e) show a dramatic reduction in salt-and-pepper noise. FTSDT-fixed appears to over-smooth some areas (e.g., blue color in the white circles of Fig. 5.10d), likely due to fixed square neighborhoods. In contrast, FTSDT-adaptive's predictions show less over-smoothing effect in the white circles. The reason is that its focal function is computed based on flexible neighborhoods adapted to the spatial topological relationship among locations. FTSDT-adaptive has somewhat lower spatial autocorrelation in predictions than FTSDT-fixed due to less aggressive smoothing, but its overall accuracy is better.

5.5.3 Computational Performance

This section compares the computational performance of the new FTSDT-Train-Refined algorithm with the baseline FTSDT-Train algorithm on different parameter settings. For simplicity, we fixed the neighborhood type as fixed neighborhoods.

5.5.3.1 Different Numbers of Training Samples N

We fixed the variables as follows: $N_0 = 50$, $S_{max} = 5$, and $N_d = 256$ and increased the number of training samples.

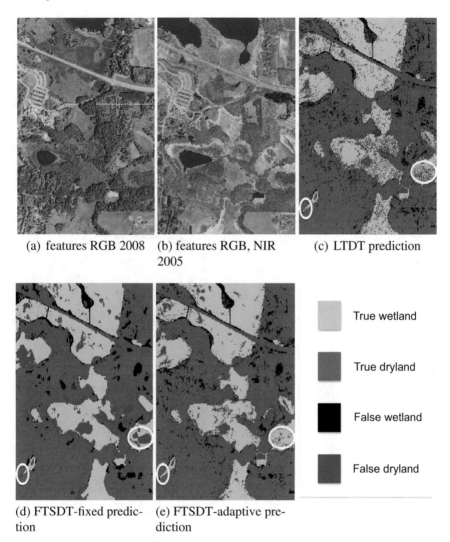

(a) features RGB 2008   (b) features RGB, NIR 2005   (c) LTDT prediction

True wetland

True dryland

False wetland

False dryland

(d) FTSDT-fixed prediction   (e) FTSDT-adaptive prediction

Fig. 5.10 Case study dataset and prediction results of LTDT and FTSDT

Figure 5.11a shows the result. As can be seen, when the training sample size is very small (e.g. 1000), the time cost of both algorithms is close. However, as the training sample size increases, the time cost of the baseline algorithm increases at a much higher rate than the refined algorithm. This result accords with cost models in *Lemmas 1* and 2, which showed that the baseline algorithm FTSDT-Train has a larger constant factor on the $O(\log N + N_d S^3)$ term.

Fig. 5.11 Computational
performance comparison
between basic and refined
algorithms

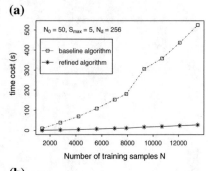

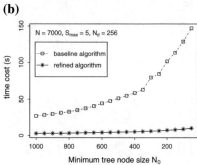

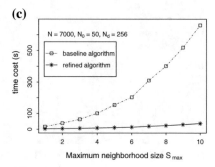

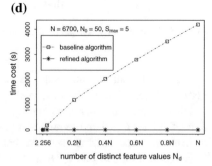

5.5.3.2   Different Minimum Tree Node Sizes N_0

We fixed the variables $N = 7000$, $S_{max} = 5$, and $N_d = 256$ and increased the minimum tree node size.

Figure 5.11b shows the result. As can be seen, as the minimum tree node size increases, the time cost of both algorithms decreases. The reason is that fewer tree nodes are constructed and thus less computation is needed. But our refined algorithm has persistently lower cost than our baseline algorithm. This result aligns with previous cost models in *Lemmas 1* and *2*, where the baseline algorithm has a larger numerator.

5.5.3.3   Different Maximum Neighborhood Sizes S_{max}

We fixed the variables $N = 7000$, $N_0 = 50$, and $N_d = 256$ and increased the maximum neighborhood size.

Figure 5.11c shows the result. As can be seen, when the maximum neighborhood size is very small (i.e., 1), the time cost of both algorithms is close, due to the low time cost when S_{max} is very small. However, as the maximum neighborhood size increases, the time cost of the baseline algorithm grows dramatically faster than the refined algorithm. This result matches the cost models in *Lemmas 1* and *2*, where the baseline algorithm has a larger constant factor N_d on the $O(N_d S^3)$ term.

5.5.3.4   Different Number of Distinct Feature Values N_d

We fixed the variables $N = 6700$, $N_0 = 50$, and $S_{max} = 5$ and increased the number of distinct feature values N_d from 2 to 256 (default value without adding simulation), and $0.2N$, $0.4N$, $0.6N$, $0.8N$ to approximately N. In order to control N_d in datasets, we independently added random noise in uniform distribution $U(0, 0.01)$ to each feature value and then specified the precision of the decimal part or even the integer part (greater precision increases N_d values). The reported N_d values are the averages across all features.

Figure 5.11d shows the result. As can be seen, when $N_d = 2$ (the first tick mark), the time cost of the two algorithms is very close (baseline algorithm costs 3.3s and refined algorithm costs 7.2s). The reason is that the focal computation cost of even the baseline approach is very small when N_d is close to 1 (baseline cost is slightly lower due to other constant factors). However, as N_d increases, the time cost of the baseline algorithm grows almost linearly to N_d while that of the refined algorithm remains the same.

This result can be explained by the cost models in *Lemmas 1* and *2*, where the baseline algorithm has a factor $O(\log N + N_d S^3)$ while the refined algorithm has a corresponding term $O(\log N + S^3)$. As the cost models imply, when N_d is close to 1 (e.g., 2), the two algorithms' time costs are very close. But as N_d increases, the cost of the baseline algorithm is a linear function of N_d given other variables are constant.

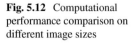

Fig. 5.12 Computational performance comparison on different image sizes

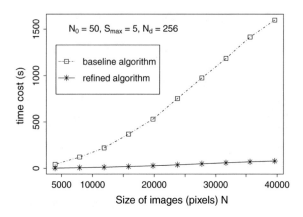

5.5.3.5 Different Image Sizes N

We fixed the variables $N_0 = 50$, $S_{max} = 5$, and $N_d = 256$ and increased the size of training image (in terms of the amount of pixels) from 3960, 7920, ..., to 39600.

Figure 5.12 shows the result. As can be seen, when the training image size is very small (e.g. 3960 pixels), the time cost of both algorithms is close. However, as the training image size increases, the time cost of the baseline algorithm increases at a much higher rate than the refined algorithm. This shows that the refined algorithm is much more scalable to large image sizes.

5.6 Discussion

A number of other relevant techniques exist for reducing salt-and-pepper noise. Examples include preprocessing (median filtering [24], weighted median filtering [29], adaptive median filtering [30], decision based filtering [31, 32]), post-processing (per parcel classification [33], spectral and spatial classifications [34]), and adding contextual variables to the input features [35]. An increasingly popular technique is image segmentation, especially geographic object-based image analysis (GEOBIA) [36]. In this technique, the image is first segmented into different objects. Then, each object is a minimum classification unit. All these techniques can help reduce salt-and-pepper noise. However, they require labor-intensive manual tuning by domain experts beyond model learning and prediction. Our FTSDT approach automates the tuning process in model learning and prediction, potentially saving users hours of labor.

5.7 Summary

This work explores the spatial decision tree learning problem for raster image classification. The problem is challenging due to the spatial autocorrelation effect and computational cost. Related work is limited to using local tests in tree nodes. In contrast, we introduce our focal-test-based spatial decision tree (FTSDT) model and its learning algorithm. We further illustrate methods for computational optimization with a refined algorithm that selectively updates focal values. Both theoretical analysis and experimental evaluation show that our refined algorithm is more scalable than our baseline algorithm. We also design a new focal test approach with adaptive neighborhoods to avoid over-smoothing in wedge-shaped areas. Experiment results on real-world datasets show that new FTSDT with adaptive neighborhoods improves classification accuracy of both the default FTSDT with fixed neighborhoods and traditional LTDT.

References

1. J.R. Quinlan, in *C4.5: Programs for Machine Learning* (Morgan kaufmann, 1993)
2. Z. Jiang, S. Shekhar, P. Mohan, J. Knight, J. Corcoran, Learning spatial decision tree for geographical classification: a summary of results, in *SIGSPATIAL/GIS* (2012), pp. 390–393
3. M.A. Friedl, C.E. Brodley, Decision tree classification of land cover from remotely sensed data. Remote Sens. Environ. **61**(3), 399–409 (1997)
4. J.M. Corcoran, J.F. Knight, A.L. Gallant, Influence of multi-source and multi-temporal remotely sensed and ancillary data on the accuracy of random forest classification of wetlands in northern minnesota. Remote Sens. **5**(7), 3212–3238 (2013)
5. J.F. Knight, B.P. Tolcser, J.M. Corcoran, L.P. Rampi, The effects of data selection and thematic detail on the accuracy of high spatial resolution wetland classifications. Photogram. Eng. Remote Sens. **79**(7), 613–623 (2013)
6. B. Walsh, How wetlands worsen climate change (2010), http://www.time.com/time/health/article/0,8599,1953751,00.html
7. A. Deschamps, D. Greenlee, T. Pultz, R. Saper, Geospatial data integration for applications in flood prediction and management in the red river basin, in *International Geoscience and Remote Sensing Symposium, Toronto, Canada. Symposium, Geomatics in the Era of RADARSAT (GER'97)* (Ottawa, Canada, 2002)
8. R. Hearne, Evolving water management institutions in the red river basin. Environ. Manag. **40**(6), 842–852 (2007). Springer
9. C. Van Westen, Remote sensing for natural disaster management. Int. Arch. Photogr. Remote Sens. **33**(B7/4; PART 7), 1609–1617 (2000)
10. A. Akselrod-Ballin, M. Galun, R. Basri, A. Brandt, M. Gomori, M. Filippi, P. Valsasina, An integrated segmentation and classification approach applied to multiple sclerosis analysis, in *2006 IEEE Computer Society Conference on Computer Vision and Pattern Recognition*, vol 1 (IEEE, 2006), pp. 1122–1129
11. M. Celebi, H. Kingravi, Y. Aslandogan, W. Stoecker, Detection of blue-white veil areas in dermoscopy images using machine learning techniques, in *Proceedings of SPIE Vol*, vol 6144 (Citeseer, 2006), pp. 61445T–1
12. D. Bazell, D.W. Aha, Ensembles of classifiers for morphological galaxy classification. Astrophys. J. **548**(1), 219 (2001)

13. T. Yuan, W. Kuo, Spatial defect pattern recognition on semiconductor wafers using model-based clustering and bayesian inference. Eur. J. Oper. Res. **190**(1), 228–240 (2008)
14. J. Quinlan, Induction of decision trees. Mach. Learn. **1**(1), 81–106 (1986). Springer
15. L. Breiman, J. Friedman, C.J. Stone, R.A. Olshen, in *Classification and Regression Trees* (Chapman & Hall/CRC, 1984)
16. X. Li, C. Claramunt, A spatial Entropy-Based decision tree for classification of geographical information, in *Transactions in GIS*, vol 10(3) (Blackwell Publishing Ltd., 2006), pp. 451–467
17. D. Stojanova, M. Ceci, A. Appice, D. Malerba, S. Džeroski, Global and local spatial autocorrelation in predictive clustering trees, in *Discovery Science* (Springer, 2011), pp. 307–322
18. D. Stojanova, M. Ceci, A. Appice, D. Malerba, S. Dzeroski, *Dealing with Spatial Autocorrelation when Learning Predictive Clustering Trees* (Elsevier, Ecological Informatics, 2012)
19. Z. Jiang, S. Shekhar, X. Zhou, J. Knight, J. Corcoran, Focal-test-based spatial decision tree learning: a summary of result, in *2013 IEEE 13th International Conference on Data Mining (ICDM)* (IEEE, 2013)
20. Z. Jiang, Learning spatial decision trees for land cover mapping, in *2015 IEEE International Conference on Data Mining Workshop (ICDMW)* (IEEE, 2015), pp. 1522–1529
21. Z. Jiang, S. Shekhar, X. Zhou, J. Knight, J. Corcoran, in *Focal-test-based Spatial Decision Tree Learning. IEEE Transactions on Knowledge and Data Engineering*, (2015)
22. A.H. Solberg, T. Taxt, A.K. Jain, A markov random field model for classification of multisource satellite imagery. IEEE Trans. Geosci. Remote Sens. **34**(1), 100–113 (1996)
23. M. Celik, B.M. Kazar, S. Shekhar, D. Boley, D.J. Lilja, Spatial dependency modeling using spatial auto-regression, in *Workshop on Geospatial Analysis and Modeling with Geoinformation Connecting Societies (GICON), International Cartography Association (ICA)* (2006)
24. C. Boncelet, Image noise models. ed. by A.C. Bovik, in *Handbook of Image and Video Processing*, Chap. 4.5., 2nd edn. (Academic Press, 2005)
25. L. Anselin, Local indicators of spatial association–lisa. Geograph. Anal. **27**(2), 93–115 (1995)
26. R.G. Congalton, A review of assessing the accuracy of classifications of remotely sensed data. Remote Sens. Environ. **37**(1), 35–46 (1991)
27. J.D. Bossler, J.R. Jensen, R.B. McMaster, C. Rizos, in *Manual of Geospatial Science and Technology* (CRC Press, 2004)
28. R.G. Congalton, K. Green, in *Assessing the Accuracy of Remotely Sensed Data: Principles and Practices* (CRC Press, 2008)
29. D. Brownrigg, The weighted median filter. Commun. ACM **27**(8), 807–818 (1984)
30. H. Hwang, R.A. Haddad, Adaptive median filters: new algorithms and results. IEEE Trans. Image Process. **4**(4), 499–502 (1995)
31. R.H. Chan, C.-W. Ho, M. Nikolova, Salt-and-pepper noise removal by median-type noise detectors and detail-preserving regularization. IEEE Trans. Image Process. **14**(10), 1479–1485 (2005)
32. S. Esakkirajan, T. Veerakumar, A.N. Subramanyam, C. PremChand, Removal of high density salt and pepper noise through modified decision based unsymmetric trimmed median filter. Sig. Process. Lett. IEEE **18**(5), 287–290 (2011)
33. J. Wijnant, T. Steenberghen, Per-parcel classification of urban ikonos imagery, in *Proceedings of 7th AGILE Conference on Geographic Information Science* (2004), pp. 447–455
34. Y. Tarabalka, J.A. Benediktsson, J. Chanussot, Spectral-spatial classification of hyperspectral imagery based on partitional clustering techniques. IEEE Trans. Geosci. Remote Sens. **47**(8), 2973–2987 (2009)
35. A. Puissant, J. Hirsch, C. Weber, The utility of texture analysis to improve per-pixel classification for high to very high spatial resolution imagery. Int. J. Remote Sens. **26**(4), 733–745 (2005)
36. G. Hay G. Castilla, Geographic object-based image analysis (geobia): A new name for a new discipline, in *Object-based Image Analysis* (Springer, Berlin, 2008), pp. 75–89

Chapter 6
Spatial Ensemble Learning

Abstract This chapter introduces a novel ensemble learning framework called spatial ensemble, which is used to classify heterogeneous spatial data with class ambiguity. Class ambiguity refers to the phenomenon whereby samples with similar features belong to different classes at different locations (e.g., spectral confusion between different thematic classes in earth observation imagery). This chapter also provides preliminary results of comparison between spatial ensemble and traditional ensemble learning (e.g., bagging, boosting, and random forest) on wetland mapping datasets.

6.1 Introduction

Classifying heterogeneous geographic data with class ambiguity, i.e., same feature values corresponding to different classes in different locations, is a fundamental challenge in machine learning [1, 2]. Figure 6.1 shows an example in a wetland mapping application. The goal is to classify remote sensing image pixels (Fig. 6.1a) into wetland and dryland classes (Fig. 6.1b). The two circled areas contain pixels that share very similar spectral values yet belong to two different classes (also called spectral confusion). As a result, a decision tree classifier learned from the entire image makes tremendous prediction errors as shown in Fig. 6.1d. The goal of spatial ensemble learning is to decompose the geographic area into zones so as to minimize class ambiguity and to learn a local model in each zone.

Motivations: Spatial ensemble learning can be used in many applications where geographic data is heterogeneous with class ambiguity. For example, in remote sensing image classification, spectral confusion is a challenging issue. The issue is particularly important in countries where the type of auxiliary data that could reduce spectral confusion, such as elevation data, or imagery of high temporal and spatial resolution, is not available. In economic study, it may happen that old house age indicates high price in rural areas but low price in urban areas [3]. Thus, *age* can be an effective coefficient to classify house price in individual zones but ineffective in a global model. In cultural study, touching somebody during conversation is welcomed in France and Italy, but considered offensive in Britain unless in a sport field; the "V-Sign" gesture can mean "two" in America, "victory" in German, but "up yours"

© Springer International Publishing AG 2017
Z. Jiang and S. Shekhar, *Spatial Big Data Science*,
DOI 10.1007/978-3-319-60195-3_6

(a) Spectral features in remote sensing image

(b) Ground truth classes (red: dry land,
green: wetland)

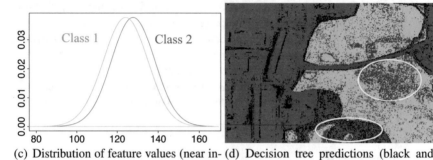

(c) Distribution of feature values (near infrared band) in circles

(d) Decision tree predictions (black and
blue show errors)

Fig. 6.1 Real-world example of heterogeneous geographic data: class ambiguity exists in two *white circles*

in Britain [4]. In these cases, spatial ensemble learning can provide a tool that captures heterogeneous relationships between factors (e.g., house age and gestures) and target phenomena (e.g., house price and culture meanings).

Challenges: The SEL problem is computationally challenging. First, there are a large number of spatial samples (pixels) to partition. Second, the objective measure of class ambiguity is non-distributive, i.e., the degree of class ambiguity in a zone cannot be easily computed from the degrees of class ambiguity in its subzones. Finally, given a geographic data, the number of candidate partitions is exponential to the number of spatial samples. It can be proved that finding an optimal zone partition is NP-hard.

Related work: Spatial ensemble learning belongs to a general category of ensemble learning problems [5–7], in which a number of weak models are combined to boost prediction accuracy. Conventional ensemble methods, including bagging [8], boosting [9], and random forest [10], assume an identical distribution of samples. Thus, they cannot address heterogeneous geographic data with class ambiguity. Decomposition-based ensemble methods (also called divide-and-conquer), including mixture of experts [11, 12] and multimodal ensemble [13], go beyond the identical and independent distribution assumption in that these methods can partition multi-modular input data and learn models in local partitions. Partitioning is usually conducted in feature vector space via a gating network, which can be learned simultaneously by an EM algorithm, or modeled by radius basis functions [14] or multiple local ellipsoids [15]. However, partitioning input data in feature vector space cannot effectively separate samples with class ambiguity because such samples are very "close" in non-spatial feature attributes. Moreover, adding spatial coordinates into feature vectors may still be ineffective since it creates geographic partitions whose zonal footprints are hard to interpret and can be too rigid to separate ambiguous zones with arbitrary shapes.

There are other techniques for heterogeneous data. A geographically weighted model [3] uses spatial kernel weighting functions to learn local models, but cannot allow arbitrary shapes of spatial zones for local models. Gaussian process [16] and multi-task learning [17] can also be used for heterogeneous geographic data, but they do not particularly focus on the class ambiguity issue. The mixture of experts approach has been used for *scene classification* on images via sub-blocks partitioning and learning local experts. But that problem is to classify an entire image (not individual pixels) [18].

To address the limitations of related work, we introduce a recently proposed spatial-based ensemble learning framework [19] that explicitly partitions input data in geographic space. Our approach first preprocesses data into homogeneous spatial patches and then uses a greedy heuristic to allocate pairs of patches with high class ambiguity into different zones.

We make the following contributions: (1) we formulate a novel spatial ensemble learning problem to classify heterogeneous geographic data with class ambiguity; (2) we propose efficient zone partitioning algorithms based on greedy heuristics; (3) we provide theoretical analysis on the proposed approach; (4) we evaluate the proposed spatial ensemble approach against related works on three real-world wetland mapping datasets.

Scope: This chapter focuses on the class ambiguity issue in heterogeneous geographic data. Other recent advances that do not address class ambiguity, such as spatial–spectral classifiers [20], object-based image analysis [21, 22], metric learning, and active learning, fall outside the scope of this work.

Outline: The chapter is organized as follows. Section 6.2 defines basic concepts and formalizes the spatial ensemble learning problem. Section 6.3 introduces our approach. Experimental evaluations are in Sect. 6.4. Section 6.5 concludes the chapter with future work.

6.2 Problem Statement

6.2.1 Basic Concepts

Geographic raster framework: A geographic raster framework $\mathbf{F}$ is a tessellation of a 2-D plane into a regular grid. Each grid cell (or pixel) is a *spatial data sample*,

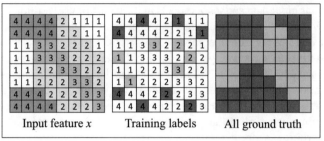

(a) Problem inputs

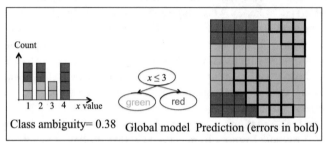

(b) Problem outputs for global model

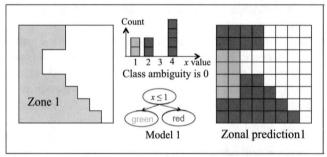

(c) Problem outputs for spatial ensemble part 1

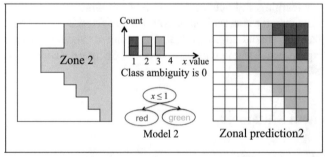

(d) Problem outputs for spatial ensemble part 2

Fig. 6.2 Illustrative example of problem inputs and outputs

defined as $s_i = (x_i, l_i, y_i)$, $1 \leq i \leq |\mathbf{F}|$, where x_i is a non-spatial feature vector, l_i is a 2-dimensional vector of spatial coordinates, and $y_i \in \{c_1, c_2, ..., c_p\}$ is a class label among p categories. All the samples in $\mathbf{F}$ can be divided into two disjoint subsets, a *labeled sample set* $\mathbf{L} = \{s_i = (x_i, l_i, y_i) \in \mathbf{F} | y_i$ is known$\}$ and *unlabeled sample set* $\mathbf{U} = \{s_i = (x_i, l_i, y_i) \in \mathbf{F} | y_i$ is unknown$\}$. In the example of Fig. 6.2a, $\mathbf{F}$ has 64 samples, including 14 labeled samples (colored in "training labels") and 50 unlabeled samples. Each sample has a 1-dimensional feature x and a class label (*red* or *green*).

A *spatial neighborhood relationship* is a Boolean function on two samples $\mathcal{R}(s_i, s_j)$, whose value is *true* if and only if s_i and s_j are adjacent (i.e., two cells share a boundary).

A *patch* $\mathbf{P}$ is a spatially contiguous subset of samples, formally, $\mathbf{P} \subseteq \mathbf{F}$ such that for any two samples $s_i, s_j \in \mathbf{P}$, either $\mathcal{R}(s_i, s_j)$ is *true* or we can find a set of samples $s_{p_1}, s_{p_2}, ..., s_{p_L} \in \mathbf{P}$ such that $\mathcal{R}(s_i, s_{p_1})$, $\mathcal{R}(s_{p_k}, s_{p_{k+1}})$, and $\mathcal{R}(s_{p_L}, s_j)$ are all *true* for $1 \leq k \leq L - 1$. For example, all samples with input feature value 3 in Fig. 6.2a form a patch.

A *zone* $\mathbf{Z}$ is a subset of samples in a raster framework $\mathbf{Z} \subseteq \mathbf{F}$. We define a function $f(\cdot)$ on a zone $\mathbf{Z}$ as the number of *isolated patches* in $\mathbf{Z}$. For example, in the left figure of Fig. 6.2a, samples with input feature value 1 form a *zone* with two isolated *patches*, while samples with feature value 3 form a *zone* with only one *patch*. Note that a zone can contain both labeled samples $\mathbf{L_Z} = \mathbf{L} \cap \mathbf{Z}$ and unlabeled samples $\mathbf{U_Z} = \mathbf{U} \cap \mathbf{Z}$ (Table 6.1).

Table 6.1 A list of symbols and descriptions

Symbol	Description
$\mathbf{F}$	All samples in a raster framework
$\mathbf{L}$	All labeled samples in $\mathbf{F}$
$\mathbf{U}$	All unlabeled samples in $\mathbf{F}$
s_i	The ith spatial data sample
x_i	The vector of non-spatial features
l_i	The vector of two spatial coordinates
y_i	The class label of sample s_i
$\mathcal{R}(s_i, s_j)$	Spatial neighborhood relationship
$\mathbf{P}$	A patch
$\mathbf{Z}$	A zone
$f(\mathbf{Z})$	The number of patches in $\mathbf{Z}$
$\mathbf{L_Z}$	All labeled samples in $\mathbf{Z}$
$N_k(s_i)$	Feature space neighborhood of s_i
$a(s_i)$	Per sample class ambiguity
$a(\mathbf{Z})$	Per zone class ambiguity

Class ambiguity refers to the phenomenon whereby samples with the same non-spatial feature vector belong to different classes, due to spatial heterogeneity (e.g., heterogeneous terrains). For example, in Fig. 6.2a, the four samples labeled with feature value $x = 1$ belong to different classes (two *red* and two *green*). A global decision tree model makes erroneous predictions (Fig. 6.2b). The degree of class ambiguity in a zone $\mathbf{Z}$ can be measured on its labeled samples $\mathbf{L_Z}$. We define the following concepts to quantify class ambiguity:

The *feature space neighborhood* of a sample s_i among all labeled samples $\mathbf{L_Z}$ in zone $\mathbf{Z}$ is defined as $N_k(s_i) = \{s_j \in \mathbf{L_Z} | s_j \neq s_i, d(x_i, x_j) \text{ is the k smallest}\}$, where $d(x_i, x_j)$ is a metric function such as Euclidean distance. We assume labeled samples are locally dense in feature space. For example, for the red sample in the last column in the middle of Fig. 6.2a, its $N_2(s_i)$ can be any two labeled samples with $x = 1$ except the sample itself, including one *red* sample and two *green* samples.

The *per sample class ambiguity* on a labeled sample s_i among all labeled sample $\mathbf{L_Z}$ in zone $\mathbf{Z}$ is defined as the ratio of labeled samples in different class from s_i in its neighborhood $N_k(s_i)$. Formally,

$$a(s_i) = \frac{1}{k} \sum_{s_j \in N_k(s_i)} I(y_j \neq y_i), \tag{6.1}$$

where $I(\cdot)$ is an indicator function. For example, the class ambiguity of the red sample in the last column of Fig. 6.2a, if one *red* sample and one *green* sample (with feature $x = 1$) are selected as $N_2(s_i)$, is $\frac{1}{2} = 0.5$. The expectation of $a(s_i)$ in this case is $2/3 = 0.67$.

The *per zone class ambiguity* of a zone is defined as the average of *per sample class ambiguity* over all labeled samples. Formally,

$$a(\mathbf{Z}) = \frac{1}{|\mathbf{L_Z}|} \sum_{s_i \in \mathbf{L_Z}} a(s_i) \tag{6.2}$$

For example, in Fig. 6.2a, b, the class ambiguity in the zone of the entire raster framework is $(\frac{2}{3} \times 4 + \frac{2}{3} \times 4 + 0 \times 2 + 0 \times 4)/14 = 0.38$. Similarly, the per zone class ambiguity of $\mathbf{Z_1}$ or $\mathbf{Z_2}$ in Fig. 6.2c is 0.

A *spatial ensemble* is a decomposition of a raster framework $\mathbf{F}$ into m disjoint zones $\{\mathbf{Z_1}, \mathbf{Z_2}, .., \mathbf{Z_m}\}$ such that the average per zone class ambiguity is minimized. For simplicity, only one local model is learned in each zone $\mathbf{Z_i}$ based on its labeled samples $\mathbf{L_{Z_i}}$ and is used to classify unlabeled samples $\mathbf{U_{Z_i}}$ in the same zone. The concept of a local model can be generalized to a set of models (e.g., bagging and boosting) in the zone. Figure 6.2c shows an example of spatial ensemble with $m = 2$.

6.2.2 Problem Definition

The spatial ensemble learning problem is formally defined as follows:
Input:

- A geographic raster framework $\mathbf{F}$: labeled samples $\mathbf{L}$ and unlabeled samples $\mathbf{U}$
- The number of zones in the spatial ensemble: m
- The parameter in feature space neighborhood: k
- The maximum number of patches in m zones: c_0

Output: A spatial ensemble such that:

$$\arg\min_{Z_1, Z_2, \ldots, Z_m} \frac{1}{m} \sum_{i=1}^{m} a(\mathbf{Z_i})$$

$$\text{subject to} \quad (1) \mathbf{Z_i} \cap \mathbf{Z_j} = \emptyset \text{ for } i \neq j, \bigcup_{i=1}^{m} \mathbf{Z_i} = \mathbf{F}$$

$$(2) \sum_{i=1}^{m} f(\mathbf{Z_i}) \leq c_0,$$

where $a(\mathbf{Z_i})$ is the per zone class ambiguity, and $f(\mathbf{Z_i})$ is the number of isolated patches.

The constraint (2) above is added for two reasons. First, it serves as a spatial regularization to avoid overfitting: in Fig. 6.2a, without a spatial constraint, we could simply output two disjoint zones, one with all *red* labeled (training) samples and the other with all *green* labeled samples (regardless of unlabeled samples). Such an output would have the minimum (zero) per zone class ambiguity on training samples but may not work well on unlabeled samples. Second, the constraint follows the first law of geography, "Everything is related to everything else, but nearby things are more relevant distant things" [23]. The constraint enforces zones to be somehow contiguous so that nearby locations are likely to follow the same local model.

We also make the following assumptions:

- This problem is a transductive learning problem. Feature vectors of unlabeled (test) samples are given.
- A pixel belongs to only one class.
- Labeled (training) samples are sufficient in number to be locally dense in feature space.
- The spatial autocorrelation effect exists, i.e., nearby locations resemble each other.

Illustrative example: Figure 6.2 shows a problem example. Inputs include a geographic data with 64 samples, 14 labeled (training) and 50 unlabeled, with one feature x and two classes (*red, green*) (Fig. 6.2a). The class ambiguity of the entire framework is $a(\mathbf{F}) = 0.38$, computed from the class histogram of training samples. A global decision tree makes prediction errors (Fig. 6.2b). In contrast, a spatial ensemble with two zones in Fig. 6.2c reduces per zone class ambiguity to zero. Predictions of local models show zero errors.

Theorem 6.1 *The spatial ensemble learning problem is NP-hard.*

Proof The proof comes from two aspects. First, our objective function of per zone class ambiguity measure is non-monotonic and non-distributive. Thus, we cannot compare one candidate zone partitioning against other candidate zone partitioning without computing their per zone class ambiguity. Second, the number of candidate partitioning is beyond polynomial to the number of samples. This can be derived from the NP-hardness of grid graph partitioning problems (i.e., if there are polynomial numbers of candidate partitioning, then grid graph partitioning can be solved in polynomial time).

6.3   Proposed Approach

In this section, we present the algorithms that address key computational challenges of spatial ensemble learning problem. To reduce the number of partition units, we preprocess the input data into homogeneous patches and use these patches (instead of samples) as partition units (Sect. 6.3.1). To efficiently evaluate a candidate zonal partition, we propose an approximation of *per zone class ambiguity* using patch pairwise class ambiguity in a zone (Sect. 6.3.2). Finally, we propose a greedy heuristic that separates the most ambiguous pair of patches into different zones by spatial proximity (Sect. 6.3.3). For simplicity, we focus on spatial ensemble learning with two zones ($m = 2$). More general cases with over two zones will be addressed in future work.

6.3.1   Preprocessing: Homogeneous Patches

A *homogeneous patch* is a patch in which all samples have similar feature values, and all labeled samples are in the same class. Real-world geographic data often exhibit homogeneous patches due to the spatial autocorrelation effect (the first law of geography [23]).

Given geographic data with all labeled and unlabeled samples, generating homogeneous patches can be considered as image segmentation [24] but with the constraint that labeled samples in the same patch belong to the same class. Algorithm 8 shows our method to generate homogeneous patches. First, each sample is initialized as a patch (step 1). The algorithm then repeatedly merges pairs of *adjacent patches* (patches with samples that are spatial neighbors) in a greedy manner. More specifically, patch pairs whose labeled samples belong to the same class (step 4) and whose samples have the smallest feature distance (step 5) are merged first (steps 8–9). Merging continues until the number of patches is reduced to a given number (n_p). In implementation, we can use a patch adjacency graph to efficiently find pairs of adjacent patches. The graph can be easily updated when two patches (nodes) are merged. To avoid small holes, a window filter can be applied before running our algorithms. The isolated patch number parameter can be determined by the size of the input dataset and the spatial scale of the spatial autocorrelation effect.

Algorithm 8 Homogeneous Patch Generation

Input:
- F: all samples in the raster framework
- $\mathcal{R}(\cdot, \cdot)$: spatial neighborhood relationship
- n_p: the number of output patches, $n \ll |F|$

Output:
- n_p patches: $\{P_1, P_2, ..., P_{n_p}\}$

1: Initialize $P_i \leftarrow s_i, i = 1..|F|$
2: **while** number of patches $> n_p$ **do**
3:    **for each** adjacent pair P_i and P_j **do**
4:       **if** L_{P_i}, L_{P_j} either empty or same class **then**
5:          $d(P_i, P_j) \leftarrow \frac{1}{|P_i||P_j|} \sum_{s_i \in P_i, s_j \in P_j} d(x_i, x_j)$
6:       **else**
7:          $d(P_i, P_j) \leftarrow +\infty$
8:       Find P_i, P_j with minimum dissimilarity
9:       Merge these two patches $P_i \leftarrow P_i \cup P_j$
10: **return** all current patches.

Figure 6.3 shows a toy example. The input geographic data contains 64 samples with one feature and two classes (*red* and *green*). Adjacent samples with the same feature value are merged into a patch. For instance, all samples with feature value 4 in the upper left corner are merged into patch A. The final output is 7 homogeneous patches (shown by different shades: A to G), which means preprocessing has reduced the number of partition units from 64 to 7.

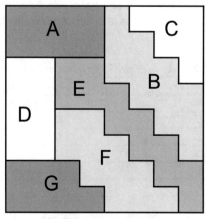

| (a) Input geographic data | (b) Homogeneous patches |

Fig. 6.3 Illustration of homogeneous patch generation

Now the problem becomes to find a partition (or *grouping*) of homogeneous patches into two disjoint sets (zones), such that the average per zone class ambiguity is minimized and the total number of *isolated patches* in zones does not exceed a threshold. Note that the number of *isolated* patches in a zone may differ from the number of *homogeneous patches*. Take Fig. 6.3 as an example. If $\mathbf{Z_1} = \{B, C, G\}$ and $\mathbf{Z_2} = \{A, D, E, F\}$, then the total number of isolated patches is actually three: two in $\mathbf{Z_1}$ ($\{B, C\}, \{G\}$) and one in $\mathbf{Z_2}$ ($\{A, D, E, F\}$).

6.3.2   Approximate Per Zone Class Ambiguity

The second challenge we address is that *per zone class ambiguity* measure in our objective is non-distributive, i.e., the degree of class ambiguity in a zone cannot be easily computed from the degrees of class ambiguity in its subzones.

Thus, we propose an *approximation* of per zone class ambiguity that can be easily computed. Gven $\mathbf{Z} = \{\mathbf{P_1}, \mathbf{P_2}, ..., \mathbf{P_{n_Z}}\}$, we define

$$\hat{a}(\mathbf{Z}) = \frac{1}{n_Z} \sum_{i=1}^{n_Z} \hat{a}(\mathbf{P_i}) = \frac{1}{n_Z} \sum_{i=1}^{n_Z} \max_{j \neq i} a(\mathbf{P_i} \cup \mathbf{P_j}), \tag{6.3}$$

where $a(\mathbf{P_i} \cup \mathbf{P_j})$ is the class ambiguity (see Eq. 6.2) over patch pair $\mathbf{P_i}$, $\mathbf{P_j}$. Our approximation $\hat{a}(\mathbf{Z})$ of per zone class ambiguity is the average of approximated class ambiguity in each patch $\hat{a}(\mathbf{P_i})$, which is defined as the maximum class ambiguity across $\mathbf{P_i}$ and any other patch. For example, in Fig. 6.3, assuming $\mathbf{Z} = \{A, B, C, D, E, F\}$, the approximated class ambiguity in each patch $\hat{a}(\mathbf{P_i})$ and in the entire zone is summarized in Table 6.2. The proposed approximation is easy to compute since it only depends on the class ambiguity of all patch pairs together.

Properties of $\hat{a}(\mathbf{Z})$: Our approximation is based on the intuition that class ambiguity within a zone comes from the class ambiguity across patches. Thus, if we can minimize the maximum class ambiguity between patches, the class ambiguity within

Table 6.2 Example of approximation in Eq. 6.3

Patch $\mathbf{P_i}$	$\mathbf{P_j}$ with max $a(\mathbf{P_i} \cup \mathbf{P_j})$	$\hat{a}(\mathbf{P_i})$
A	All the same	0
B	F	0.67
C	D	0.67
D	C	0.67
E	All the same	0
F	B	0.67
G	All the same	0
$\hat{a}(\mathbf{Z}) = \frac{0 \times 3 + 0.67 \times 4}{7} = 0.383$, where $\mathbf{Z} = \{A, ..., G\}$		

a zone can be reduced as well. We assume that labeled (training) samples in homogeneous patches $\mathbf{P_i}$ and $\mathbf{P_j}$ are sufficient (locally dense in feature space) to compute $a(\mathbf{P_i} \cup \mathbf{P_j})$. Patches without labeled samples do not contribute to the approximation measure.

6.3.3 Group Homogeneous Patches into Zones

Since class ambiguity in a zone is approximated by maximum class ambiguity across all patch pairs, we propose a greedy heuristic that separates the most ambiguous pair of patches into two zones in each iteration. The heuristic is summarized in Algorithm 9. It has three phases: allocation of ambiguous patch pairs (steps 1–11), allocation of remaining patches (step 12), and spatial adjustment to satisfy the contiguity constraint (steps 13–15).

Algorithm 9 Group Patches into Zones

Input:
  - A set of homogeneous patches: $\{\mathbf{P_1}, \mathbf{P_2}, ..., \mathbf{P_{n_p}}\}$
  - Max number of isolated patches in zones: c_0

Output:
  - Two disjoint subsets (zones) of patches: $\mathbf{Z_1}, \mathbf{Z_2}$

1: Initialize $\mathbf{Z_1} \leftarrow \emptyset, \mathbf{Z_2} \leftarrow \emptyset$
2: Compute $a_{i,j} = a(\mathbf{P_i} \cup \mathbf{P_j})$ for any pair $i \neq j$
3: Generate a set of ambiguous patch pairs:
    $\mathbf{A} = \{\{\mathbf{P_i}, \mathbf{P_j}\} | \forall i \neq j, a_{i,j} > 0\}$
4: **while** $\mathbf{A} \neq \emptyset$ **do**
5:    Find $\{\mathbf{P_i}, \mathbf{P_j}\} \in \mathbf{A}$ with the highest $a_{i,j}$
6:    **if** $\mathbf{P_i}$ or $\mathbf{P_j}$ is not yet in zones $\mathbf{Z_1}$ and $\mathbf{Z_2}$ **then**
7:      **if** only one of $\mathbf{P_i}, \mathbf{P_j}$ in zone $\mathbf{Z_1}$ or $\mathbf{Z_2}$ **then**
8:        Allocate unassigned patch to different zone
9:      **else if** neither $\mathbf{P_i}$ nor $\mathbf{P_j}$ in two zones **then**
10:       Allocate $\mathbf{P_i}, \mathbf{P_j}$ to $\mathbf{Z_1}$ or $\mathbf{Z_2}$ one-to-one separately by spatial proximity //*details in text
11:   Remove $\{\mathbf{P_i}, \mathbf{P_j}\}$ from set $\mathbf{A}$
12: Allocate remaining patches to zones by spatial proximity //*details in text
13: **while** $f(\mathbf{Z_1}) + f(\mathbf{Z_2}) > c_0$ **do**
14:   Switch the patch across zones such that $f(\mathbf{Z_1}) + f(\mathbf{Z_2})$ decreases most //*details in text
15: **return** $\mathbf{Z_1}, \mathbf{Z_2}$

Steps 1 to 3 initialize two empty sets (zones), compute class ambiguity across all patch pairs, and create a set of ambiguous patch pairs $\mathbf{A}$. $a_{i,j}$ will be zero for patches without labeled samples. Step 5 identifies the most ambiguous pair of patches from $\mathbf{A}$. Steps 6 to 10 allocate the two members of this pair to two zones. If one patch is already allocated to a zone, then the algorithm allocates the other patch to the other zone (step 8). Otherwise, if neither patch is allocated yet, the two patches are allocated to two zones one-to-one by spatial proximity (step 10). More specifically, a patch is allocated to the zone whose existing patches are closer by Euclidean distance. After a pair of patches is allocated, the patches are removed from **A**. Allocation process continues until $\mathbf{A} = \emptyset$ (all pairs of ambiguous patches are allocated to zones).

Step 12 allocates the patches that remain (patches without labeled samples and patches with no class ambiguity with others). More specifically, a remaining patch adjacent to a zone is first allocated to the zone. When multiple remaining patches are adjacent to a zone, the patch that helps balance the class frequency of the labeled samples in the zone is allocated first.

Steps 13–14 spatially adjust patch allocations in two zones to satisfy the spatial constraint. This is needed when the two zones generated from previous steps contain more than c_0 isolated patches. More specifically, each time, the algorithm switches the patch whose allocation change across zones decreases the total number of isolated patches by the most.

Running example: Fig. 6.4 shows a running example of Algorithm 9. The input data is the same as the example in Fig. 6.3, and $c_0 = 2$. Two zones (bold solid circles and regular solid circles) are initially empty. First, ambiguous patch pair $\{C, D\}$ is allocated (Fig. 6.4a). Then, $\{B, F\}$ is allocated, with B allocated to the same zone as C due to spatial proximity (Fig. 6.4b). After that, the algorithm allocates remaining patches A, E, G, respectively (Fig. 6.4c–e). Since the current total number of isolated patches is 2 ($\{B, C\}$, $\{A, D, E, F, G\}$), no spatial adjustment is needed.

6.3.4   Theoretical Analysis

Theorem 6.2 *The time complexity of homogeneous patch generation algorithm is $O((n - n_p)n^2)$ where n is the total number of samples and n_p is the number of patches.*

Proof In this algorithm, each sample is considered as a separate patch initially, resulting in n patches. In each run of the while loop at line 2, a pair of patches is merged, so the number of patches reduces by 1. Because the stop condition is the number of patches reaches n_p, this while loop runs $n - n_p$ times. In each run, all patch pairs are checked to select the most homogeneous one. There may be at most n^2 pairs. Thus, the time complexity of this algorithm with input sample size n is $O((n - n_p)n^2)$.

Theorem 6.3 *The time complexity of patch grouping is $O(n_p^2 n_l^2)$ where n_p is the number of patches, n_l is the number of labeled samples in each patch.*

Proof For simplicity, let's set n_p as the number of patches, n_l as the number of labeled samples in each patch. Pair-wise class ambiguity measure is calculated firstly. Now that there are n_p patches, and there may be at most n_p^2 patch pairs. The time complexity of generating k nearest neighbor of each labeled sample is $O(n_l^2)$. Thus, pair-wise class ambiguity calculation is $O(n_p^2 n_l^2)$. Seed Assignment part scans the ambiguity pairs whose number is at most n_p^2. Seed growing scans the unassigned patches after

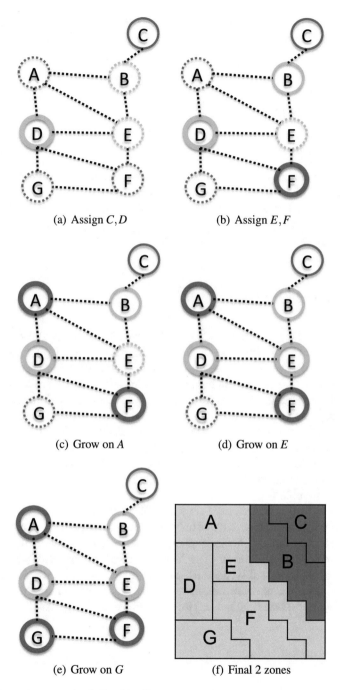

(a) Assign C, D

(b) Assign E, F

(c) Grow on A

(d) Grow on E

(e) Grow on G

(f) Final 2 zones

Fig. 6.4 A running example of Algorithm 9

seed assignment whose number is at most n_p. Therefore, the time complexity of this algorithm is dominated by pair-wise class ambiguity measure computation and is $O(n_p^2 n_l^2)$.

6.4   Experimental Evaluation

The goal of the experiments was to:

- Compare spatial ensemble learning with other conventional ensemble methods.
- Test the sensitivity of spatial ensemble to parameters.
- Compare spatial ensemble with adding spatial coordinate features in other ensemble methods (in appendix).
- Interpret results in case studies (in appendix).

6.4.1   Experiment Setup

We compared spatial ensemble with bagging, boosting, random forest, and feature vector space ensemble (mixture of experts). Bagging, boosting, and random forest were from the Weka toolbox [25]. A hierarchical mixture of experts MATLAB package with logistic regression base models [26] was used. Controlled parameters included the number of base classifiers m, the base classifier type, the size of training (labeled) samples, and the maximum total number of isolated patches in spatial ensemble c_0. We tested two more internal parameters for spatial ensemble, i.e., the number of homogeneous patches n_p in preprocessing, and the k value in class ambiguity measure (Eqs. 6.1, 6.2). In all experiments, we fixed $m = 100$ for bagging, boosting, and random forest, and $m = 2$ (one model in each zone), $c_0 = 10, k = 10$ for spatial ensemble.

Dataset description: Our datasets were collected from Chanhassen, MN [27]. Explanatory features were four spectral bands (red, green, blue, and near-infrared) in high-resolution (3m by 3m) aerial photos from the National Agricultural Imagery Program during leaf-off season. Class labels (wetland and dry land) were collected from the updated National Wetland Inventory. Within study area, we picked a scene, randomly selected a number of training (labeled) samples, and used the remaining samples (whose classes were hidden) for prediction and testing (details in Table 6.3).

Table 6.3 Dataset description

Scene	Total samples		Training set	
	Dry	Wet	Dry	Wet
Chanhassen	47077	35577	1758	2434

Evaluation metric: We evaluated the classification performance with confusion matrices and the F-score of the wetland class (wetland class is of more interest).

6.4.2 Classification Performance Comparison

6.4.2.1 Comparison on Classification Accuracy

Parameters: The base classifiers were decision trees. $n_p = 200$ for Chanhassen dataset, and $n_p = 1000$ for the other two datasets.

Analysis of results: The classification accuracy results are summarized in Table 6.4. In the confusion matrix displayed in each table, the first and second rows show *true* dryland and wetland samples, respectively, and the first and second columns show *predicted* dryland and wetland samples, respectively. We can see that bagging, boosting, and random forest reduce the number of false wetland errors (upper right) and false dryland errors (lower left) in decision tree predictions by less than 10% (e.g., from 6276 to 5939 for bagging in Table 6.4), while spatial ensemble reduces those errors by over 30% to 50% (e.g., from 6276 to 2421 for spatial ensemble in Table 6.4). This is also shown in the F-score columns.

6.4.2.2 Effect of Base Classifier Type

The parameters were the same as Sect. 4.2.1. The Chanhassen dataset was used. Base classifiers tested included decision tree (DT), SVM, neural network (NN), and logistic regression (LR). Mixture of expert was compared with the others only on logistic regression due to availability of package. Results are shown in Fig. 6.5. The spatial ensemble approach consistently outperformed the other methods on all base classifier types.

Table 6.4 Performance on Chanhassen data

Ensemble method	Confusion matrix		F-score
Single model	38,567	**6136**	0.82
	6276	27,483	
Bagging	39,298	**5405**	0.83
	5939	27,820	
Boosting	38,579	**6124**	0.83
	5653	28,106	
Random forest	39,258	**5445**	0.84
	5262	28,497	
Spatial ensemble	40,732	**3971**	0.91
	2421	31,338	

Fig. 6.5 Effect of the base
classifier type

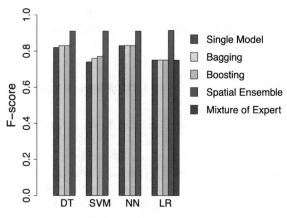

Fig. 6.6 Effect of the
number of training samples

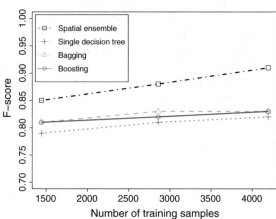

6.4.2.3   Effect of Training Set Size

We used the Chanhassen dataset and varied the number of training samples as 1444, 2857, to 4192, corresponding to 50, 100, and 150 circular clusters on class maps, respectively. The other parameters were the same as those in Sect. 4.2.1. Results are summarized in Fig. 6.6. Spatial ensemble consistently outperformed the other methods. Experiments on several more training sample sizes are needed in future work to observe how fast different methods reach best accuracy.

6.4.2.4   Sensitivity of Spatial Ensemble to n_p

We used the Chanhassen dataset and the same parameter settings as Sect. 4.2.1, except that we varied the number of patches in the preprocessing steps from 200 to 600. Results in Fig. 6.7 show that the performance of spatial ensemble approach was

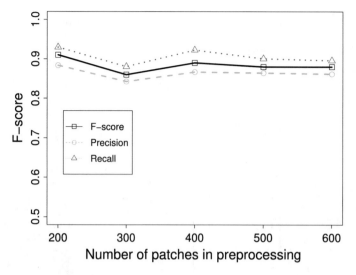

Fig. 6.7 Effect of n_p in preprocessing

generally stable, with slightly lower accuracy when the number of patches was 300, but the performance in all cases was better than bagging and boosting (F-score below 0.83).

6.4.3 Effect of Adding Spatial Coordinate Features

Given the same problem input as Fig. 6.2a in the chapter, simply adding spatial coordinates into feature vectors and then running a global model or random forest is ineffective (Fig. 6.8). The reason is that this approach is sensitive to training sample locations and may be insufficient to address the arbitrary shapes of local zones.

In our experiment, we also investigated if adding spatial coordinates in feature vectors will always be effective in reducing class ambiguity. We used the Chanhassen data. The parameter settings were the same as Sect. 4.B.1 except that the training set size was smaller (624 wetland samples and 820 dryland samples, within 50 small circular clusters). Training sample locations were shown in Fig. 6.9a, where almost all training samples on the left half belonged to the dryland class (red). Due to this reason, decision tree and random forest models mistakenly "learned" that almost all samples in the left half should be predicted as dryland class (red). Thus, we can see that parts of wetland parcels in the left half of the image were misclassified (black errors in Fig. 6.9b, c). Mixture of experts approach also made similar mistakes (black errors in Fig. 6.9d), though the errors were slightly less serious. In contrast, spatial ensemble did not have same misclassification due to its more flexible spatial partition.

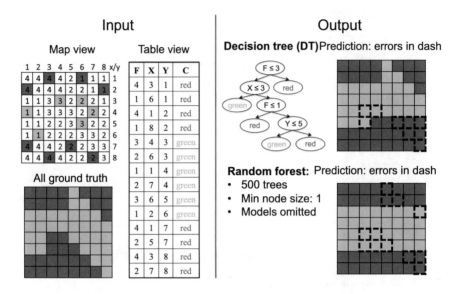

Fig. 6.8 Illustrative example of adding spatial coordinate features

The experiment showed that adding spatial coordinates in feature vectors in related work may not always be sufficient, particularly when sample locations are too sparse to capture the footprint shapes of class patches.

6.4.4   Case Studies

Figure 6.10 shows a case study for Chanhassen, MN, and the results of the spatial ensemble approach were interpreted by domain experts in remote sensing and wetland mapping. The datasets and parameter configurations were the same as those in Sect. 4.2.1. The input spectral image features, ground truth wetland class map, as well as output predictions from a single decision tree and spatial ensemble (SE) were all shown in the figure, numbered by different study areas.

The study area in general shows a good spectral separability for the SE prediction results (Fig. 6.10b) between true dry land representing "red" that is uplands land cover and true wetland represented as "green" for wetlands land cover. On the other hand, there was higher spectral confusion when the decision tree prediction (Fig. 6.10c) was used compared to the SE prediction results. This spectral confusion can be explained primarily because of the different types of wetland and upland features found in these areas. For example, for the Chanhassen data (Fig. 6.10a), two main different features were found as the main cause of spectral confusion: tree canopy vs. forested wetlands; these two features have different physical

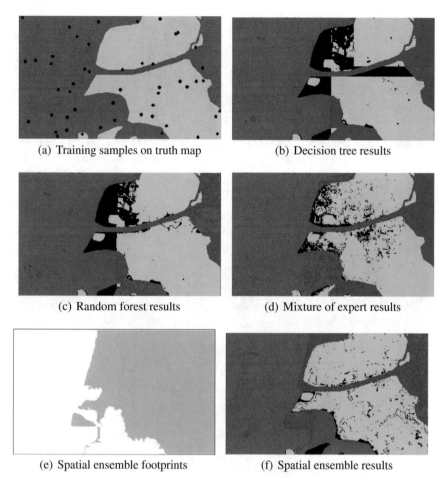

(a) Training samples on truth map   (b) Decision tree results

(c) Random forest results   (d) Mixture of expert results

(e) Spatial ensemble footprints   (f) Spatial ensemble results

Fig. 6.9 Comparison with related work adding spatial coordinate features in decision tree, random forests, and mixture of experts (*black* and *blue* are errors, best viewed in color)

characteristics but similar spectral properties in the image data. This makes difficult to discriminate because a forested type of wetlands will appear cover with vegetation in the aerial imagery but in the real world, it is very different compared to the regular tree canopy feature. Spatial ensemble footprints (Fig. 6.10d) separated ambiguous areas into different local decision tree models, so there was less spectral confusion in each local model.

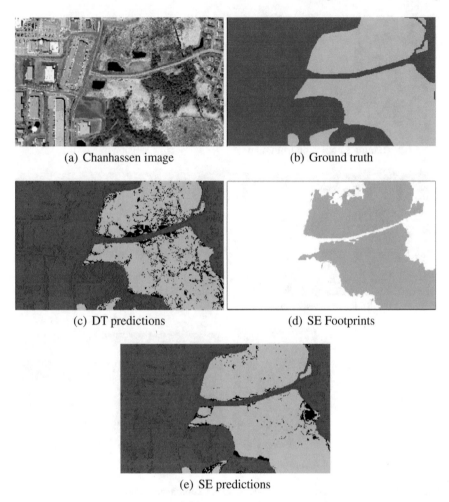

(a) Chanhassen image   (b) Ground truth

(c) DT predictions   (d) SE Footprints

(e) SE predictions

Fig. 6.10 A real-world case study in Chanhassen (errors are in *black* and *blue*)

6.5 Summary

This chapter investigates the spatial ensemble learning problem for geographic data with class ambiguity. The problem is important in applications such as land cover classification from heterogeneous earth imagery with class ambiguity, but is challenging due to computational costs. We introduce our spatial ensemble framework that first preprocesses data into homogeneous patches and then uses a greedy heuristic to separate pairs of patches with high class ambiguity. Evaluations on real-world dataset show that our spatial ensemble approach is promising. However, several issues need to be further addressed in future work. First, current spatial ensemble learning algorithms only learn a decomposition of space into two zones (for two local models). In real-world scenario, we often need multiple zones or local models. We need to generalize our spatial ensemble algorithms for more than two zones (e.g., hierarchical spatial ensembles). Second, the theoretical properties (e.g., optimality) of proposed spatial ensemble learning framework need to be further investigated. Finally, we also need to address the case with limited ground truth training labels, particularly when test samples are from a different spatial framework.

References

1. T.K. Ho, M. Basu, Complexity measures of supervised classification problems. IEEE Trans. Pattern Anal. Mach. Intel. **24**(3), 289–300 (2002)
2. T.K. Ho, M. Basu, M.H.C. Law, Measures of geometrical complexity in classification problems, in *Data Complexity in Pattern Recognition* (Springer, 2006), pp. 1–23
3. A.S. Fotheringham, C. Brunsdon, M. Charlton, in *Geographically Weighted Regression: The Analysis of Spatially Varying Relationships* (Wiley, 2003)
4. B. Pease, A. Pease, in *The Definitive Book of Body Language* (Bantam, 2006)
5. T.G. Dietterich, Ensemble methods in machine learning, in *Multiple Classifier Systems* (Springer, 2000), pp. 1–15
6. Z.-H. Zhou, *Ensemble Methods: Foundations and Algorithms* (CRC Press, 2012)
7. Y. Ren, L. Zhang, P. Suganthan, Ensemble classification and regression-recent developments, applications and future directions [review article]. Comput. Intel. Mag. IEEE **11**(1), 41–53 (2016)
8. L. Breiman, Bagging predictors. Mach. Learn. **24**(2), 123–140 (1996)
9. Y. Freund, R.E. Schapire, A decision-theoretic generalization of on-line learning and an application to boosting. J. Comput. Syst. Sci. **55**(1), 119–139 (1997)
10. L. Breiman, Random forests. Mach. Learn. **45**(1), 5–32 (2001)
11. R.A. Jacobs, M.I. Jordan, S.J. Nowlan, G.E. Hinton, Adaptive mixtures of local experts. Neural Comput. **3**(1), 79–87 (1991)
12. S.E. Yuksel, J.N. Wilson, P.D. Gader, Twenty years of mixture of experts. IEEE Trans. Neural Netw. Learn. Syst. **23**(8), 1177–1193 (2012)
13. A. Karpatne, A. Khandelwal, V. Kumar, Ensemble learning methods for binary classification with multi-modality within the classes, in *Proceedings of the 2015 SIAM International Conference on Data Mining, Vancouver, BC, Canada, April 30 - May 2, 2015* (SIAM, 2015), pp. 730–738
14. L. Xu, M.I. Jordan, G.E. Hinton, An alternative model for mixtures of experts, in *Advances in Neural Information Processing Systems* (1995), pp. 633–640
15. V. Ramamurti, J. Ghosh, Advances in using hierarchical mixture of experts for signal classification, in *1996 IEEE International Conference on Acoustics, Speech, and Signal Processing, 1996. ICASSP-96. Conference Proceedings*, vol. 6 (IEEE, 1996), pp. 3569–3572
16. G. Jun, J. Ghosh, Semisupervised learning of hyperspectral data with unknown land-cover classes. IEEE Trans. Geosci. Rem. Sens. **51**(1), 273–282 (2013)
17. A.R. Gonçalves, F.J. Von Zuben, A. Banerjee, Multi-label structure learning with ising model selection, in *Proceedings of the 24th International Conference on Artificial Intelligence* (AAAI Press, 2015), pp. 3525–3531
18. M. Szummer, R.W. Picard, Indoor-outdoor image classification, in *Proceedings of the 1998 IEEE International Workshop on Content-Based Access of Image and Video Database, 1998* (IEEE, 1998), pp. 42–51
19. Z. Jiang, S. Shekhar, A. Kamzin, J. Knight, Learning a spatial ensemble of classifiers for raster classification: A summary of results, in *2014 IEEE International Conference on Data Mining Workshop (ICDMW)* (IEEE, 2014), pp. 15–18

20. M. Fauvel, Y. Tarabalka, J.A. Benediktsson, J. Chanussot, J.C. Tilton, Advances in spectral-spatial classification of hyperspectral images. Proc. IEEE **101**(3), 652–675 (2013)
21. D. Lu, Q. Weng, A survey of image classification methods and techniques for improving classification performance. Int. J. Rem. Sens. **28**(5), 823–870 (2007)
22. J. Dong, W. Xia, Q. Chen, J. Feng, Z. Huang, S. Yan, Subcategory-aware object classification, in *Proceedings of the IEEE Conference on Computer Vision and Pattern Recognition* (2013), pp. 827–834
23. W.R. Tobler, A computer movie simulating urban growth in the detroit region. Econ. Geogr. **46**, 234–240 (1970)
24. R.M. Haralick, L.G. Shapiro, Image segmentation techniques, in *1985 Technical Symposium East*. International Society for Optics and Photonics (1985), pp. 2–9
25. Weka 3: Data mining software in java (2016), http://www.cs.waikato.ac.nz/ml/weka/
26. D.R. Martin, C.C. Fowlkes, Matlab codes for multi-class hierarchical mixture of experts model (2002), http://www.ics.uci.edu/~fowlkes/software/hme/
27. L.P. Rampi, J.F. Knight, K.C. Pelletier, Wetland mapping in the upper midwest united states. Photogram. Eng. Rem. Sens. **80**(5), 439–448 (2014)

Part III
Future Research Needs

Chapter 7
Future Research Needs

Abstract This chapter summarizes the future research needs and concludes the book.

7.1 Future Research Needs

Modeling complex spatial dependency: Most current research in spatial classification uses Euclidean space, which often assumes isotropic property and symmetric neighborhoods. However, in many real-world applications, the underlying spatial dependency across pixels shows a network topological structure, such as pixels on river networks. One of the main challenges is to account for the network structure in the dataset. The network structure often violates the isotropic property and symmetry of neighborhoods, and instead, requires asymmetric neighborhood and directionality of neighborhood relationship (e.g., network flow direction based on elevation map). Recently, some cutting edge research in spatial network statistics and data mining [1] has proposed new statistical methods such as a network K-function and network spatial autocorrelation. Several spatial analysis methods have also been generalized to network space, including a network point cluster analysis and clumping method, network point density estimation, network spatial interpolation (Kriging), as well as a network Huff model. Due to the nature of distinct spatial network space compared to Euclidean space, these statistics and analysis often rely on advanced spatial network computational techniques [1]. The main issue is to incorporate asymmetric spatial dependency into classification models and to develop efficient learning algorithms.

Mining heterogeneous earth observation imagery: The vast majority of existing earth imagery classification algorithms assume that data distribution is homogeneous. In other words, models learned from a training pixels can be applied to other test pixels. However, as we previously discussed, spatial data is heterogeneous in nature. Applying a global model to every subregion may have poor performance. In Chap. 6, we introduce some preliminary work on spatial ensemble learning, which decomposes study area into different homogeneous zones and learns local models

© Springer International Publishing AG 2017
Z. Jiang and S. Shekhar, *Spatial Big Data Science*,
DOI 10.1007/978-3-319-60195-3_7

for different zones. However, it remains a challenge on how to effectively and efficiently identify a good spatial decomposition. The main challenge lies in the fact that we need to consider both the geographical space and feature-class space when evaluating a candidate decomposition. To reducing computational cost, top-down hierarchical spatial partitioning may be a potential solution. The advantage of top-down hierarchical spatial partitioning is that we can stop partitioning a subregion if the data distribution inside the region is already homogeneous, and focus more on the subregions where data distribution is non-homogeneous, e.g., with class ambiguity. Another important issue related to spatial heterogeneity is on how to adapt a model learned from one region to a new region with a different data distribution. This is similar to transfer learning in machine learning community. However, unique spatial characteristics such as spatial autocorrelation and dependency have to be incorporated.

Fusing earth imagery big data from diverse sources: Earth imagery data comes from different sources (e.g., different satellite and airborne platforms). These diverse imagery have different spatial, spectral, and temporal resolutions. For example, MODIS imagery provides global coverage around every other day, but the spatial resolution is coarse (around 200 m). Landsat imagery, however, has higher spatial resolution (30 m), but relatively less frequent in temporal dimension. Most of existing classification methods work on a single type of earth imagery data. Since each data source itself is not perfect with its own data quality issues such as noise and cloud, classification algorithms that can utilize multiple data imagery data sources all together to address the data quality issues in each single data source are of significant practical values. Fusioning these diverse imagery data, however, is challenging due to different spatial, temporal, and spectral resolutions, spatial coregistration issues, and requirements of novel model learning algorithms. Moreover, earth observation imagery can also be integrated with data in other modality such as text data in geosocial media, and in situ group point samples. Multi-view learning can be a potential solution due to its capability to integrate features of different types.

Spatial big data classification algorithms: Another important future research direction is to develop scalable spatial big data classification algorithms. In applications where the spatial scale is large (e.g., global or nation wide study) or the spatial resolution is very high (e.g., precision agriculture), the size of data can be enormous, exceeding the capability of a single computational device. In these situations, developing scale algorithms on big data platforms is very important. Due to the unique data characteristics of spatial data such as spatial autocorrelation, anisotropy, and heterogeneity, parallelizing spatial classification algorithms is often more challenging than traditional non-spatial (or per-pixel) classification algorithms. Due to the fact that spatial classification algorithms are often computationally intensive (sometimes involving large number of iterations), GPU and Spark platforms are more appropriate. Future work is needed to characterize the computational structure of different spatial classification algorithms to design parallel version on big data platforms. For example, the learning algorithms of focal-test-based spatial decision tree in Chap. 5 involve a large number of neighborhood statistics computation, which can be easily parallelized in GPU devices. However, the computation of focal function values across different candidates has a dependency structure due to the incremental update method, making it hard to parallelize. Future work is needed to address these challenges.

7.2 Summary

We introduce spatial big data and overview current spatial big data analytic algorithms for different tasks. We particularly focus on novel spatial classification techniques for earth observation imagery big data. Earth imagery big data is important for many applications such as water resource management, precision agriculture, and disaster management. However, classifying earth imagery big data poses unique computational challenges such as spatial autocorrelation, anisotropy, and heterogeneity. We introduce several examples of recent spatial classification algorithms including spatial decision trees and spatial ensemble learning that address some of these challenges. We also discuss some remaining challenges and potential future research directions.

Reference

1. A. Okabe, K. Sugihara, *Spatial Analysis Along Networks: Statistical and Computational Methods*. (Wiley, 2012)